CELL BIOLOGY - NEW INSIGHTS

Edited by **Stevo Najman**

Cell Biology - New Insights

http://dx.doi.org/10.5772/59649
Edited by Stevo Najman

Contributors

Stevo J Najman, Sanja Stojanović, Liora Shoshani, Teresa Del Carmen López-Murillo, Omar Paez-Gómez, María Luisa Roldán, Claudia Andrea Vilchis-Nestor, Jorge Alberto Lobato-Alvarez, Danuta Radzioch, Laura Lasagni, Duccio Lombardi, Juan Carlos Gallego-Gomez, Juan Carlos Cardona, Leandro Ariza-Jimenez, Elizabeth Orozco-García

CBS, Edition **2017**

Published by InTech

Janeza Trdine 9, 51000 Rijeka, Croatia

Publishing Process Manager Iva Lipovic

Technical Editor InTech DTP team

Cover Designer InTech Design Team

Additional hard copies can be obtained from orders@intechopen.com

Cell Biology - New Insights, Edited by Stevo Najman
p. cm.
ISBN 978-953-51-2242-5

INTECH

open science | open minds

Contents

Contents

Preface

Cell biology is a multidisciplinary scientific field that its modern expansion in new knowledge and applications owes to important support of new technologies with the rapid development, such as ICTs. By integrating knowledge from nano-, molecular, micro-, and macroareas, it represents a strong foundation for almost all biological sciences and disciplines, as well as for biomedical research and application. This book is a compilation of inspiring reviews/original studies, which are divided into sections: New Methods in Cell Biology, Molecular and Cellular Regulatory Mechanisms, and Cellular Basis of Disease and Therapy. The book will be very useful for students and beginners to gain insight into new area, as well as for experts and scientists to find new facts and expand their scientific horizons through biological sciences and biomedicine.

Stevo Najman, PhD
Resident Professor of Biology and Genetics
Head of Department of Biology and Human Genetics
Head of Scientific-Research Center for Biomedicine
Faculty of Medicine, University of Niš

Part-time Professor at Department of Biology and Ecology
Faculty of Science, University of Niš

Section 1

New Methods in Cell Biology

Chapter 1

A Proposal for a Machine Learning Classifier for Viral Infection in Living Cells Based on Mitochondrial Distribution

Juan Carlos Cardona-Gomez, Leandro Fabio Ariza-Jimenez and Juan Carlos Gallego-Gomez

Additional information is available at the end of the chapter

http://dx.doi.org/10.5772/61293

Abstract

The study of viral infections using live cell imaging (LCI) is an important area with multiple opportunities for new developments in computational cell biology. Here, this point is illustrated by the analysis of the sub-cellular distribution of mitochondrium in cell cultures infected by Dengue virus (DENV) and in uninfected cell cultures (Mock-infections). Several videos were recorded from the overnight experiments performed in a confocal microscopy of spinning disk. The density distribution of mitochondrium around the nuclei as a function of time and space $\rho(r, \theta, t)$ was numerically modeled as a smooth interpolation function from the image data and used in further analysis. A graphical study shows that the behavior of the mitochondrial density is substantially different when the infection is present. The DENV-infected cells show a more diffuse distribution and a stronger angular variation on it. This behavior can be quantified by using some usual image processing descriptors called *entropy* and *uniformity*. Interestingly, the marked difference found in the mitochondria density distribution for mock and for infected cell is present in every frame and not an evidence of time dependence was found, which indicate that from the start of the infections the cells are showing an altered subcellular pattern in mitochondrium distribution. Ulteriorly, it would be important to study by analysis of time series for clearing if there is some tendency or approximate cycles. Those findings are suggesting that using the image descriptors *entropy* and *uniformity* it is possible to create a machine learning classifier that could recognize if a single selected cell in a culture has been infected or not.

Keywords: Computational Cell Biology, Dengue Virus, Mitochondria, Machine Learning

1. Introduction

At the latter years of the past century, cell biology experienced a fast growth, thanks to the convergence of several techniques, which have substantially improved the confocal microscopy field. Now, the observation in real time of the structural and functional unit of life is possible. The ultrarefrigerated CCD-cameras with electromultipliers; the implementation of confocality based on disk spinning without the necessity of high-energy lasers (which could damage the living cells in a few seconds); the increasing capacity of computational processors; and the ability of the genetic engineering for coding fluorescent proteins mutants [1, 2], offering the possibility of a color palette that was previously unthinkable for the cell molecular biologists [3, 4]. All together with the ability to generate cells with fluorescent compartments, opened the doors to maybe the most remarkable and important scientific and technological development for a new era in cell biology named Live Cell Imaging. Before the 1990'-s this kind of research was known by the unpopular name "time-lapse video microscopy", as it is detailed in a protocol book widely known at that time written by A.J. Lacey [5].

At the beginning of the new millennium, the necessity of introducing new and improved mathematical and computational tools was made evident. This was because the amount of data produced in a single experiment could overload the capacity of personal computers and the conventional software was not loaded with the required algorithms to process such data. Then a strategic alliance with researchers on the areas of artificial intelligence, applied mathematics, and physics was apparent. These new cooperations make perfect sense due that even from the beginning of life science studies, it was clear that the dynamical rules involved were complex, non-linear, and possibly not even deterministic but probabilistic. The virologists, for example, have discovered that the infection rate is governed by a non-linear pattern and the cellular physiology of several processes turn out to be more complicated than it was expected. In consequence, the mathematical modeling became the main strategy in the journey for knowing and understanding the cell biology. The amount of the data available nowadays could not be analyzed by conventional human heuristics. Fortunately, the computational biology field and its tools offer the required resolution and robustness in diverse problems. It goes even further, because the computational algorithms work evenly in any case. When dealing with complex biological problems, to have a working computational model will get us closer to the reality and help us avoiding the human bias present in heuristic approaches. It is in complex problems where the convenience of using powerful statistical tools to build models became apparent. The main strategy here is to try to "learn" the model directly from the experimental or observational data.

The term "machine learning" (ML) refers to a branch of the artificial intelligence field, that concerns to the study and construction of algorithms with the ability to learn from the existing data. In such algorithms, a set of parameters is fitted to provide the best input-output relationship between the information available. When talking about a computational code that implements the techniques, algorithms, or principles found in machine learning theory, it is usually called a machine learning program. The literature on this topic is quite large; however, some very popular books are those by Duda et al. [6], Webb [7], and Bishop [8].

A commonly accepted definition of the process of "learning" is due to Tom M. Mitchell [9]: *a machine learns to perform a task T if its performance as measured by $\mathcal{P}$ increases with the experience E*. The experience *E* is the feedback the machine received to validate its output. The ML set of techniques have a broad range of applications in several fields of knowledge, including the building of autonomous robots [10], the astrophysical data mining [11], the study of dynamical systems and complex networks without the explicit knowledge of the dynamical equations [12], the patterns and shape recognition [6, 8] in images as used by the face recognition programs in social networks web pages, the hand writing OCR (optical character recognition), and of course, in medicine (automatic diagnosis based on symptoms) and biology (gene sequencer, classification of cellular morphology, etc.).

Particularly, the shape recognition capabilities have important applications to live cell image processing. For example, in 2006, Neumann and collaborators use live cell imaging to study the RNAi screening [13]. They developed a ML that recognize the morphologies present in the cells images and associate them (classify) with the corresponding phenomenology: interphase, mitosis, apoptosis and binucleated cells morphologies were studied through a multi-class classifier using support vector machine (SVM). They reported to obtain up to 97% accuracy from the SVM in comparison with "manual" classification through the observation by some very well-trained biologists.

Due to the huge amount of data provided by a live cell image (LCI) measure, it becomes unpractical to relay only on the lecture and interpretation by a well-trained researcher. It is also possible to have different interpretations coming from different scientists when analyzing the same image. Then, the ML ability to recognize and characterize particular morphologies present in an image is very useful to avoid the slow and tedious process of visual discrimination. Also, it can avoid some human bias by following well-defined rules. However, to fully train the machine, it can be necessary to have a large number of image samples from the phenotype under study. It means, to have enough sample cells expressing the phenotype and some other cells to use as a control group. Sometimes, that condition is not fulfilled. In order to asset this kind of problem, Thouis R. Jones and collaborators implement a ML with interactive feedback to characterize diverse and complex morphological phenotypes [14]. They use the criteria of well-trained researchers as a feedback in the learning stage of the machine, and provide the code [15] for the world to use under a free license.

Several generic implementations of ML techniques have been developed and presented as toolbox in scientific software. However, it is pretty common to find the particular phenomenon under study to be better fitted by some unique implementation developed explicitly to deal with it. This can be a consequence of the particularities of the problem or sometimes this is just due to the lack of proper documentation on the available tools.

This chapter is organized as follows: first, a brief description of some common methods used to build a ML are provided, followed by a description of the performed experiment and computational analysis to obtain the information from the graphical data. Finally, the results and a proposal to create a ML to characterize viral infection are presented.

2. Machine learning concepts overview

From the mathematical point of view, a ML can have one of two primary objectives: regression and classification. When the machine is used to compute the best response to a given situation among a continuous range of possible answers, it is called a regression problem. And when the machine is due to choose among a discrete set of possibilities, it is called a classification problem.

The shape recognition and feature extraction from images is a classification problem, where the duty of the machine is to find the class which has the highest probability to contain the current input value. In this context, a class is defined based on a set of measurable attributes found in an image; it can be geometrical attributes (length, shape, eccentricity, size), pixel intensity, etc. In general, the input for a ML program is a set of measurable variables or attributes, which are set in vectors. It is common in ML literature to call these attributes *features* and the vectors *feature vectors*, so these names will be used in such framework in the rest of this chapter. Each input in a feature vector represents an attribute and each vector represents a state of the system.

Before the machine is ready to be used as a classifier or predictor, it needs to be trained among some data. Here, "training" refers to the process of parameters optimization, where the machine is optimized to get the best result against the training set. This process is not perfect, and some human criteria need to be implemented. If the model has not enough freedom to fit to the training set, it gets under fitted and do not reproduce the characteristics of the system under study. On the other hand, having too much freedom in the ML leads to a model that fits pretty well in the training set, but is unable to predict accurately the outcome for a feature vector outside of the training set. This is called bias. To avoid bias, it is customary to split the available data in two sets, the training set and the testing set. A trained machine is challenged with the testing set, and the accepted ML model is the one that has the best results against it.

There are two main paradigms for the training of a classifier, the supervised and the unsupervised learning. In supervised learning, each sample in the training set is consisting on a features vector and a class flag, i.e., the classes to which each sample in training set belongs are known a priori. After the learning process, a computational model that can predict the right class flag for most of the training set is obtained, and hopefully, it would predict the correct class for a new sample with high accuracy. Also, the classifier must return information about how confident its prediction is, i.e., a value of dude must be reported.

When no predefined classification is available, a ML algorithm can be used to search for common patterns or similarities into the training set, which do not contain class information yet. The machine would cluster samples with similar feature vectors to define a class, and then, it will use the found class to characterize new input data. To do that, it is necessary to define some measure of similarity (Euclidean distance in features space, for example) that can be used to group the input vectors into clusters. The objectives of this kind of ML are first to cluster the data from the training set into classes, and then, set a classifier to characterize new inputs.

ML are suitable to treat complex problems in which the explicit mathematical form describing the interactions occurring in the process are not known, i.e., the dynamical equation ruling the systems are unknown. Being so, the computations involved in a model built with ML are not deterministic but probabilistic, based on the information gathered by direct measures. The more data are available to train the machine, the more accurate the prediction will become.

2.1. Supervised learning

The objective of the classifier is to draw a frontier that splits the feature space into k disjoint subsets [16] called the border line or border hyperplane. Hopefully, each subset will contain the feature subspace associated with one single class.

2.1.1. Hypothesis function

Let the features space be called $\mathcal{X}$, where a feature vector is x, and let the whole set of classes be $\mathcal{Y}$, with an individual class denoted y. Develop a ML model consists in building a function $h\ :\mathcal{X}\rightarrow\mathcal{Y}$ that is a good predictor of the corresponding y to a given input x. In supervised learning, a computational model function with adaptable parameters θ, $h_\theta(x)$must be build; this function is usually called "hypothesis". Also, a cost function $J(\theta)$ must be defined to provide an idea of how accurate the hypothesis is when predicting the output values or classes for the whole training set. A common choice is the least square cost function:

$$J(\theta)=\frac{1}{2}\sum_{i=1}^{m}\left(h_\theta\left(x^{(i)}\right)-y^{(i)}\right)^2, \tag{1}$$

where m is the number of samples or elements in the training set, and the notation $x^{(i)}$ and $y^{(i)}$ denotes the feature vector of the i-th sample and the class value for it. The optimization of the model (learning process) is then achieved by minimizing the cost function. Once the ML model is set up, the predictions are made with the optimized hypothesis as $y\ =\ h(x)$. Depending on if $h(x)$ is a continuous function or a discrete one, the machine will be doing regression or classification respectively.

Another approach comes from a probabilistic interpretation of the hypothesis function. Suppose that $y^{(i)}=h(x^{(i)})+\in^{(i)}$, where $\in^{(i)}$ is a random error which takes care of unmodeled effects and possibly random noise. And assume that $\in^{(i)}$ are IID (independent and identically distributed), and follows a Normal distribution (Gaussian distribution), of mean zero and some variance σ^2, $P(\in^{(i)})=\frac{1}{\sqrt{2\pi\sigma}}\exp\left(\frac{-(^{(i)})^2}{2\sigma^2}\right)$, then the probability of $y^{(i)}$ conditioned to $x^{(i)}$ and parametrized by θ is:

$$P\left(y^{(i)}\mid x^{(i)};\theta\right)=\frac{1}{\sqrt{2\pi\sigma}}\exp\left(\frac{-\left(h_\theta\left(x^{(i)}\right)-y^{(i)}\right)^2}{2\sigma^2}\right). \tag{2}$$

As any set $(x^{(i)}, y^{(i)})$ is independent from the others, the probability of the whole set $P(\mathcal{Y} \mid \mathcal{X};\theta)$ is the product of all the individual probabilities. When $P(\mathcal{Y} \mid \mathcal{X};\theta)$ is taken as a function of the parameters θ, it is called the likelihood function:

$$\ell(\theta) = \prod_{i=1}^{m} P\left(y^{(i)} \mid x^{(i)};\theta\right). \tag{3}$$

The principle of maximum likelihood establishes that the best model representation of the data is given by the set of parameters θ that provides the maximum probability. Then, maximizing the likelihood function for the whole training set (or any monotonically increasing function of it) is equivalent to minimize the cost function Eq. (1).

2.1.2. *Logistic regression*

Suppose the problem at hand is to determine if the measure of some experiment belong to one out of two possible outputs (like, for example, to determine if a tumor is benign or malign). A class flag 0 or 1 must be associated for each output. In this case, a common approach is to propose a logistic function (also known as sigmoid function) as a classifier, it is called a *logistic regression*:

$$g\left(\theta^T x\right) = \text{sig}\left(\theta^T x\right) = \frac{1}{1+\exp\left(-\theta^T x\right)}, \tag{4}$$

$$h_\theta(x) = g\left(\theta^T x\right) \tag{5}$$

The sigmoid function sig(z) has asymptotic values of 1 when $z \to \infty$ and 0 when $z \to -\infty$. So the conditional probability for the feature vector x to belong to each one of the two available classes is written as:

$$\begin{aligned} P(y=1 \mid x;\theta) &= h_\theta(x), \\ P(y=0 \mid x;\theta) &= 1-h_\theta(x). \end{aligned} \tag{6}$$

Which can be summarized in a single probability density function (PDF)

$$P(y \mid x;\theta) = \left(h_\theta(x)\right)^y \left(1-h_\theta(x)\right)^{1-y}. \tag{7}$$

Once the PDF is set, the process of learning consist in maximizing the likelihood of such PDF to the training data set. By computational simplicity, it is convenient to maximize instead some monotonically increasing function of the likelihood. It is common to work with the logarithm of the likelihood (log-likelihood function). When using the logistic regression, this hypothesis function would not return the prediction of an output class, but the probability for the sample feature vector belongs to a given class.

If it is needed to get a class value as an output, it can be done by setting $z = \mathrm{sig}(\theta^T x)$ instead of Eq. (4) and defining

$$g(\theta^T x) = \begin{cases} 1 & \text{if } z > 0.5, \\ 0 & \text{otherwise,} \end{cases} \tag{8}$$

then minimizing the cost function Eq. (1). This last strategy is known as the *perceptron learning algorithm*. The classifier in this example can be extended to k classes by a simple one vs all algorithm. It is, defining a logistic regression to compute the probability for any of the k classes, $h_{\theta_1,1}(x), h_{\theta_2,2}(x), h_{\theta_{k-1},k-1}(x), \ldots, h_{\theta_k,k}(x)$, and returning the class which has the highest probability. Note that for each class, a parameters vector θ_i must be optimized.

2.1.3. *Non-linear classifiers*

One limitation of the techniques summarized so far is that they provide a linear model for the classifier, i.e., the decision border is represented by a straight line or hyperplane. This can work perfectly fine if the data are linear separable, or if a linear border line provides enough accuracy in the final prediction. What happens if the feature space requires a more complex non-linear decision border? i.e., if the decision border is given considerably better by an hyper surface? One possible way to create non-linear models is to use *neural networks* (NN). A neuron is a computational unit, i.e., a piece of code that performs a single task or function, usually called the activation function. This method was developed as a gross mimic of a biological neuron, where each computational neuron has a set of wires connecting it with its input and a set of wires that are used to communicate its output to the next set of neurons. Each wire between two neurons has associated a parameter, sometimes called "weight", which is adjusted in the leaning process. So the computational model is created as an array of neurons, configured in layers that can be fully or partially connected. This kind of computational structure allows the creation of pretty complex nonlinear functions just by the selection of the network wires. After optimization, the NN computes a continuous function ready to be used for regression. For classification, a helper selection function can be implemented. A simple and yet powerful arrangement is the feedforward structure (see Figure 1), in which the neurons are arranged in a network where each layer receives its inputs directly from the layer before, and provides its outputs only to the layer after it, i.e., where the i-th layer receives the information from the (i - 1)-th layer, and sends its output to be the inputs to the (i + 1)-th layer. A common neural network used for classifiers is a three layer fully connected network, i.e., each neuron receives information for any neuron in the layer before. The first layer, called input layer, has a computational unit for each attribute in the feature vector. This unit sends its associated value to any neuron in the second or hidden layer, where each neuron computes a sigmoid function

$h_{\theta_i}(x)=g(\theta_i^T x)$ as in Eq. (4). θ_i is the parameters vector, containing the weights for each input wire to neuron i. Each neuron in the hidden layer sends its output to any neuron in the third or output layer. And finally, the neurons in the output layer computes another logistic function $h(\theta_j'^T z)$, where z is the vector of outputs from the second layer and θ_j' is the vector of weights for each input wire to the output neuron j. $h(\theta_j'^T z)$ corresponds to the probability for the input feature vector to belong to the class j. Then, a selector function chooses the class with the highest probability to be assigned as the final output of the classifier.

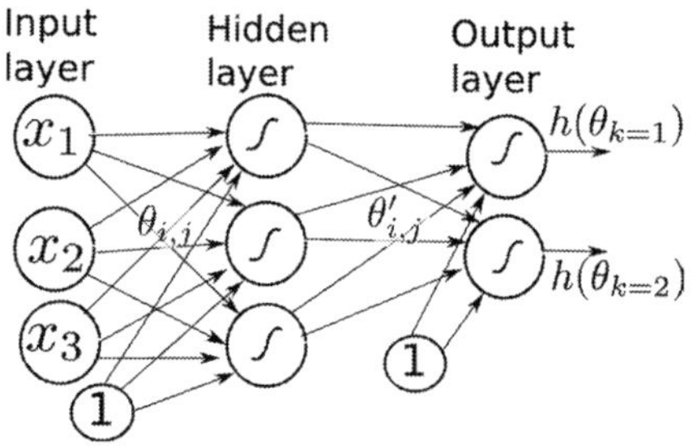

Figure 1. A schematic figure of a feed-forward three layers neural network. Here, an input vector of three features is classified in one out of two known classes. The hidden and output layers compute logistic functions in each neuron, represented by the s-like curve. $\theta_{i,j}$ is the weight of the i-th wire to the neuron j.

Any neuron in the second and third layer has an associated vector of parameters that need to be trained. The training of a NN is a difficult task, where the weights connecting each pair of neuron must be learned for all neuron in any layer. For the output layer, the cost function can be computed taking into account the expected values in the training set. But for the inner layers, no expected value is known. As a consequence of this, the cost function associated to a NN is in general a non-convex function. This has strong repercussions in the optimization problem. Due to the existence of several local minimum, the convergence to a global minimum is not guaranteed.

The variational parameters have several ways to be changed that will provide approximately the same level of correction from one iteration to the next. One strategy to find the "right direction" to move the network is to take minimal changes, i.e., from all the possible variation of parameters providing the same level of correction, the network is changed in the way that the set of parameters defers the less from its previous state. This is done by applying a generalization of the gradient descent method to deal with multilayer networks, called the back propagation algorithm, the complete description of which will be found elsewhere [17].

2.1.4. *Support vector machine*

Another method to create a nonlinear classifier is the *support vector machine* SVM. To illustrate the idea behind SVM, consider a two class problem. The main objective is to draw the decision border between two classes that provides the best separation of sub sets among all the possible border lines (see Figure 2).

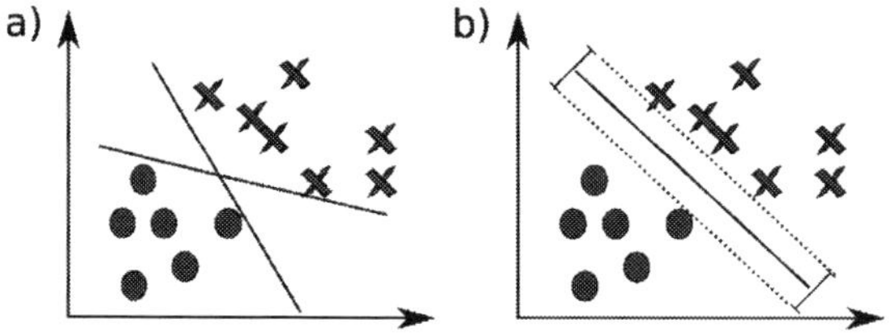

Figure 2. Among all the possible border lines, SVM chooses the one that provides the maximum margin between the classes. Here, two classes, marked with a circle and an x in a two-dimensional feature space are shown. a) Some possible non-optimal border lines. b) The right, the maximum margin border. Margins are shown in dots.

SVM finds the border line that has the largest margin or distance from the closest sample of each class. The equation of the decision border is then the equation of the hyperplane:

$$W^T x + b = 0 \tag{9}$$

W is a vector perpendicular to the hyperplane. x is a vector in feature space and b is a bias term. In this part, it is convenient to split the optimization parameters θ in W and b, to explicitly take the interception term apart from others. Now propose the hypothesis function:

$$h(x) = g\left(W^T x + b\right); \tag{10}$$

$$g(z) = \begin{cases} 1 \, if \, z \geq 1, \\ -1 \, if \, z \leq 1, \end{cases} \tag{11}$$

and the optimization problem is to find the optimal values for W and b that maximize the margin size. Starting from the Lagrangian

$$\mathcal{L}(W,b,\alpha) = \frac{1}{2} W^T W - \sum_{i=1}^{m} \alpha_i \left[y^{(i)} \left(W^T x^{(i)} + b \right) - 1 \right], \tag{12}$$

which includes the Lagrange multipliers α_i to hold the restrictions imposed by $g(z)$ in Eq. (11). The optimality conditions for the objective function are found by differentiating Eq. (12) against W and b and setting it equal to zero. It can be found that $W = \sum \alpha_i y^{(i)} x^{(i)}$ and

$\sum_{i}^{m} \alpha_i y^{(i)} = 0$. Also, the multipliers vanish for all the feature vectors outside the margin lines, the remaining vectors are called support vectors and give rise to the method's name. After some algebra, the minimization problems turn out to be the maximization of the objective function [17]

$$\mathcal{J}(\alpha) = \sum_{i}^{m} \alpha_i - \frac{1}{2} \sum_{i}^{m} \sum_{j}^{m} \alpha_i \alpha_j y^{(i)} y^{(j)} K\left(x^{(i)}, x^{(j)}\right), \tag{13}$$

under the constrains $\alpha_i \geq 0$ and $\sum_{i}^{m} \alpha_i y^{(i)} = 0$. Here, the kernel function is just the inner product $K(x^{(i)}, x^{(i)}) = (x^{(i)})^T x^{(i)}$ that corresponds to a linear classifier, i.e., when the two classes can be separated by a straight line. In the case that the data are not linear separable, SVM can become non-linear by simply replacing the kernel by a non-linear one. If the kernel represents the inner product of two vectors in the feature space, what does it mean a non-linear kernel function? When changing the kernel, a representation of the feature space in a higher dimensional space is obtained, related to the original feature space by some nonlinear transformation than is not necessary to know. The only thing required is the form of the inner product in the new coordinates expressed in terms of the original ones. This "kernel trick" allows the computation of pretty complex decision borders. However, not any function of two features constitutes a valid kernel. To solve this situation, a special case of Mercer's theorem [18] guarantees the validity of a kernel function, as far as the kernel matrix $k_{i,j} = K(x^{(i)}, x^{(j)})$ is a symmetric positive semi-definitive matrix.

In real application, it is common to find that the training set at hand is not separable, i.e., it is not possible to find a border hyper surface that splits the feature space without misclassification of some training samples. And forcing the model to fit any training vector will produce high bias. Then it is a good practice to implement *regularization*. This is done by introducing the regularization term $C \sum_{i}^{n} \xi([\theta]_i)$ in the cost to minimize. Here $[\theta]_i$ is the i-th component of the parameter vector, ξ is a penalty function which accomplish $\xi(\theta) \geq 0$ and C is the regularization parameter set by the user. C allows to control how much bias is acceptable for the final model.

2.1.5. *Confusion matrix*

To assess the accuracy of the ML model a *confusion matrix* for our classifier can be build. The confusion matrix is an evaluation of how many feature vectors in the training and/or testing set are misclassified. The matrix is built by contrasting the predicted class flag with the real one for each feature vector. For example, consider a two class problem (like the benign or malign tumor problem) that when trained against a set of 100 feature vectors report the classification presented in table 1:

Predicted / Real	Benign	Malign
Benign	45	12
Malign	15	25

Table 1. Example of a confusion matrix for a two classes clasiffier based on imaginary benign or malign tumor data.

In the example, 45 benign tumors have been classified as benign by the ML, and 15 have been misclassified as malign tumors. The error estimate for a ML code is computed as the average of miscalculated classes over the total of samples. It is the sum over all the off diagonal elements over the total number of cases. In the example, the error range of 30%. And the accuracy, defined as 1 − *err* is 70%.

2.2. Unsupervised learning

In this case, the first objective of the ML is to find some similarities among the data, which can be used to divide it into clusters. Each cluster will then define a class, and new inputs to the machine (outside the training set) will be classified following the clustering of the training space.

2.2.1. *Clustering*

So far, the feature vector is represented as a set of numerical values in real space. From the mathematical point of view, each input "box" inside a vector is called a dimension and the value in the box is a coordinate. The length of the vector is the number of dimensions of the containing space. A space is a collection of vectors that follow a set of rules (an algebra). If a distance measure for any two points in the space (features vectors) can be build, then the data can be clustered [19]. A common choice is the Euclidean distance, defined as the square root of the sums of the squares of the differences between the coordinates of the two vectors in each dimension, i.e., if **A**, **B** are vectors in $\mathbb{R}^n$, the Euclidean distance between them is given by:

$$\|A - B\| = \sqrt{\sum_{i=1}^{n}\left(x_{A,i} - x_{B,i}\right)^2} \tag{14}$$

2.2.2. *K-means*

A common clustering algorithm known as K-means is as follows: chooses a number of clusters k to be found, and initialize the clusters centroids μ_1, $\mu_2,\ldots,\mu_k$ randomly. Then assign each one of the training feature vectors to a cluster by relating it to the closest centroid (the one which has the minimum distance to the sample). When all vectors in the training set are labeled, recompute the position of the centroids as the average of the feature vectors inside the cluster, for each cluster. If there are centroids without any feature vector assigned, it can be dismissed or repositioned randomly. Now iterate again relabeling the training set and recomputing the centroids positions until convergence is achieved.

2.2.3. *Hierarchical clustering*

It is a quite expensive algorithm to obtain clusters which is based on finding a partition hierarchy among the data. It can be started by making each feature vector a cluster with one single member. Then, the distance between any pair of vectors is computed. If the distance is lesser than some selection parameter, the clusters are mixed to form one. In the new distribution of clusters, each cluster is represented by its centroid, and iterates the process until some convergence criteria is achieved. For example, a predefined number of clusters is reached. Due to the computational cost involved in hierarchical clustering, it is not recommended for problems with large training set.

2.2.4. *The CURE algorithm*

This is a large-scale-clustering algorithm. When centroids are used to define clusters, it is expected that any cluster would have a regular shape in features space, and the space is expected to be Euclidean. The clustering using representatives (CURE) algorithm is a little more general, due that it can handle irregular shaped clusters. This method defines a cluster in terms of a set of representative members of the cluster. These representatives must be chosen in a way that they are as far as possible from each other. Then, the representatives are points on the "surface" of the cluster. This kind of construction allows any shape for the cluster, including rings. To apply the CURE algorithm, first an initial clustering must be done, then the representatives are chosen for each cluster, and finally, two clusters are united if they have a pair of representatives that are close enough following some user-definedcriteria.

3. LCI experiments and data extraction

3.1. Experiment description

3.1.1. *Materials and methods*

3.1.1.1. *Cell lines expressing fluorescent mithondria (Vero-Mito)*

The Vero epithelial cells (ATCC) were maintained under standard culture conditions as described in other works from this lab [20-21], and in another chapter of this book [22]. The temperature was set at 37°C in a humidified atmosphere of 95% air and 5% carbon dioxide. The monoclonal cell line over-expressing the plasmid pmKate2-Mito (Evrogen®) were obtained with a cell sorter (Moflo XDP, Beckman Coulter®), with ulterior antibiotic selection (Kanamycin) of transfectants during 21 days in accordance to the experimental procedures described in detail in [23].

3.1.1.2. *Virus preparation, titration and infection protocols*

The strain New Guinea of Dengue Virus Serotype 2 (DENV-2) was grown and maintained in insect cells C6/36 HT under the standard practices as described in [20, 21], and this book [22].

Briefly the DENV were amplified at a very low MOI (multiplicity of infection) to avoid genetic drift and apparition of DIs (defective interfering particles), which could be altering the whole data concerning the real synchronized infections [24]. Viral titers were detected by plaque assay, using a Vero cell monolayer culture under 1% methylcellulose overlay medium as it was reported by [20, 21]. The viral infections were done by the same way of our previously reported works [20, 21], with the difference that for live cell imaging the cells were seed and registered in 35-mm glass bottom dishes (MatTek Corporation) with 0.7 mm in thickness of the glass coverslides, which is adequate in refraction index for this kind of inverted confocal microscope for registering living cells. The negative controls of infections were named mock infections, as it had been standardized traditionally for the virology community [24].

3.1.1.3. Live cell imaging

The Vero-Mito (3x105 cells) cell line was seed in Petri dishes adequate for living cells with bottom with coverslide of 0.17 mm, and previous to the register the normal culture medium used was changed by a DMEM without red phenol for avoiding the autofluorescence of this pH-indicator chemical. The videos of living cells over expressing fluorescent mitochondria (+/- infections) were obtained with a confocal microscopy based on disk spinning Unit (Olympus®IX-81 DSU), coupled to incubator and mixing gases Tokai-Hit Co® systems, which regulate the micro environment of cell culture with temperature and carbon dioxide in all system. The mock infections and infections of the overnight micrographs were captured in an OrcaR$_2$CCD (Hamamatsu®) ultra-refrigerated camera with electro multiplier, coupled to the illumination systems with Arc burners of 150 W constituted by mercury-xenon or xenon lamps (Olympus®-MT10 Illumination System). The photonic signals emitted by the biological specimen were transduced to electromagnetic waves for the CCD (charge coupled device), transmitted by a light fiber 2 m of single quartz to the Workstation Xcellence-Pro (Olympus®) for image processing, which also include the application of deconvolution tools for improving the signal/noise ratio of images.

3.2. Image segmentation and extraction of information

A total of nine videos generated by LCI where studied. Four of them correspond to mock cells (uninfected) and five to infected cells. Each video has 36 frames taken each 20 min for a period of 12 h. The videos are recorded in color at 1024x1344ppi resolution. As the color has no relevant information, they have been converted into gray scale. The last 34 vertical lines of each frame were dismissed in order to get rid of the microscope watermark. Then the resolution has been decreased to 495 x 672. The cells in this study have been selected under the following criteria:

- The whole cell is present in any frame of the video and the nucleus was clearly distinguishable on any frame.
- There were not other cells too close, so the mitochondrial distribution does not seem to be overlapped with those from neighbor cells.

Under this procedure, 11 cells were selected, 4 mock and 7 infected. Any selected cells were modeled as having an elliptical nuclei, by manually choosing four points on the nuclei borderline and applying the Hough transform [25] (see Figure 3 and 4). The nucleus and the cell are proven to be approximately aligned [26] so the nuclear envelope is approximated by another ellipsis centered in the nuclei, with the same inclination and the axis twice as long. This rudimentary model provides us with the necessary segmentation to perform the cell tracking by simply creating a mask over the region where the ellipsis is located for any frame in the video. The images are stored as intensity matrices where each entry is a pixel. Masks are stored as matrices of the same dimension, whose entries take values of zero or one. If the pixel belongs to the segmented region the associated mask value is one. Those matrices are stored in binary format to be processed in a Python script that takes advantage of numpy and Scipy libraries for further analysis.

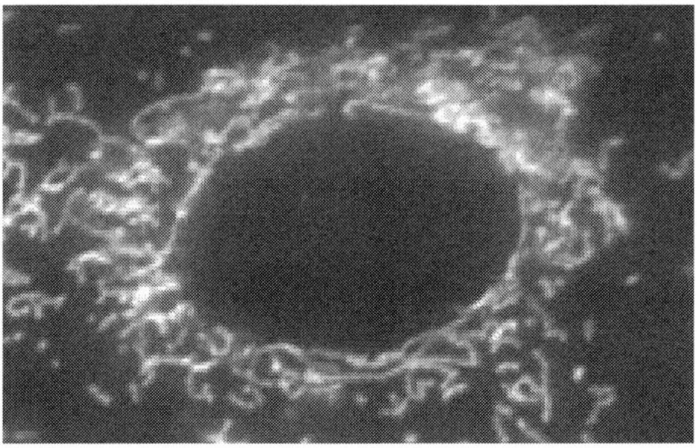

Figure 3. An infected cell appearing in the original image present on video.

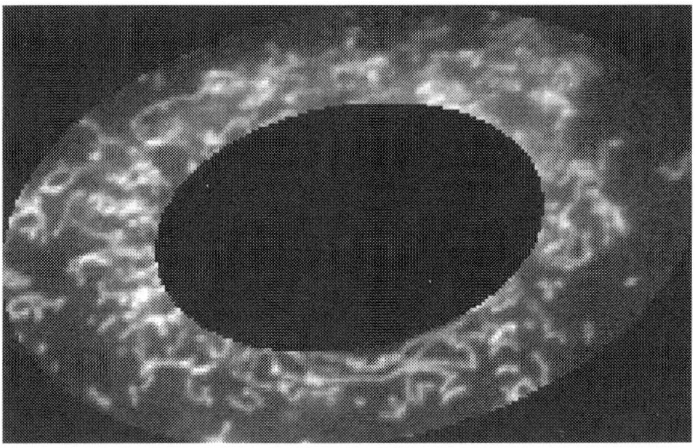

Figure 4. The segmented region used to track and study the cell. Nucleus is modeled as having an elliptic shape. The exterior membrane is modeled as a concentric ellipsis with the major semi axis twice as long as the nucleus.

http://dx.doi.org/10.5772/61293

3.3. Information processing

In the gray-scale video, the mitochondrial distribution is shown as bright points, been more brilliant those places where the density of mitochondrium is higher. Then, the density distribution of mitochondrium can be estimated as proportional to the intensity distribution $\rho(x, y, t) = \alpha I(x, y, t)$. In each frame of each video, a discrete set of pixel intensity values is recorded. This means that the intensity distribution in a discrete grid of point (X, Y) has been measured, where $x \in X$ and $y \in Y$ are the sets of pixels coordinates.

The shape of the density distribution of mitochondrium is the same shape of the intensity distribution. Those functions differ only by a constant of proportionality that became irrelevant when the density function is normalized. So further in this reading both functions would be referred indistinguishable as $\rho(x, y, t)$.

The continuous density function can be approximated from the set of pixel intensity measurements by some interpolation method. In this work, a two-dimensional interpolation in terms of bivariate splines has been used on each segmented frame. This procedure allows us to extract important information about the mitochondrial behavior. Each frame is taken after a fixed period of time of 20 mins, so they form a time series of the density distribution function $\rho(x, y, t)$ through the whole experiment.

4. Results and discussion

In cell biology studies live cell imaging is a newcomer; however, the innovation in computational biology tools is been forced by the convergence of distinct research programs. Being so, LCI is no longer only a "technique" but a new exploratory science [27], that brings the possibility of encompassing cross-disciplines. In this sense, the subcellular patterns of distinct cellular organelles and macromolecular structures within the cell are important for dynamical studies, which will be useful in predictive medicine [28].

The mitochondrial morphology is a remarkable area for biomedical research since more than a decade [29], because these cellular organelles change under physiological and pathological conditions, like metabolism, thermogenesis, homeostasis of calcium and several kinds of cell death [30, 31]. But there is lacking information about the subcellular distribution of mitochondrium after a cell injury like the viral infections, and this quantitative information is key for tracking some cellular events of virus cycle that have been covered to the computational cell biology exploration.

New developments have been focused in the high-resolution microscopy images of the fine morphology of mitochondria [32]. Having in mind the improved time resolution, this information is decisive for understanding of the dynamics and functioning of these cellular organelles at high-throughput screenings [33].

But here, the work was mainly directed to study and characterize the subcellular distribution of mitochondria with and without Dengue virus infections on epithelial cells that are constitutively expressing these organelles in red fluorescence.

Recently, it had demonstrated that not only shape, number and size of the organelles are important for the cellular function, but also their subcellular distribution, which is the consequence of the intracellular transport [34]. Since Dengue virus like many other members of the most diverse viral families are using the cytoskeleton [22], here we have tried to follow indirectly the infection process using the alterations of subcellular distributions of mitochondria.

Both mock and DENGV infected cells have been prepared by a standardized procedure that provides approximately the same initial state among all cells of each type even when different experiments are considered. So at frame 0 each of the studied cells provides a possible initial state. Each of these states must adjust to the density distribution of mitochondrium. Assuming that any possible initial state is equally probable, then our approximation to the density function at time zero $\rho(r, \theta, t = 0)$ must come from the average over all the known (measured) possible states. Each frame has the same time spacing for all the videos (20 minutes), so the same analysis is valid for the i–th frame and the mitochondria density distribution at time t would be the average over all the known states for all cells of the same kind.

The distribution as a function of two variables given by the pixel position related to the center of the cell is $\rho(x, y)$. However, before taking the average, it has to be taken into account that a single cell can have any random orientation. The elliptical shape of the nucleus makes the distribution not symmetrical. This means that all function must be rotated to have the same orientation before been able to average over them.

A Python script was written to automatically determine the major semi axis of each cell in each frame by measuring the maximum distance between two points into the segmented region. The center of the ellipsis is found as the average of all the coordinates in the image. To get $\rho(r, \theta, t)$ for any cell referred to the same polar axis, and so be able to compute the density average, the polar axis is set equal to the major semi axis in each of the cells analyzed. In polar coordinates $\rho(r, \theta)$ will give us the density of mitochondrium at a distance r from the nuclei center and at an angle θ from the major semi axis.

The average density of mitochondrium distribution $\rho(r, \theta, t)$ is shown for three frames ($t = 0$, 320, 720 minutes) in Figures 5, 6, and 7. Those are the initial state, one intermediate state and the final state of the study. The vertical axis is the level of intensity (proportional to the mitochondrium density) and the horizontal axis is the distance to the nuclei center in pixels. In each subplot the projection of $\rho(r, \theta, t)$ for a given angle in radians is presented.

It can be seen that mock cells present prominent peaks for some radial positions, i.e., the distance between the peak and the closest local minimum is large compared with the background density. Also, the background density is low. This suggests that mock cells have the mitochondrium distributed in clusters around the nuclei. At variance, the infected cells average mitochondrium density distribution presents a higher background intensity compared with the maximum of the distribution. The peaks are less defined than those for mock cells, which means that the mitochondrium clusters are less defined or inexistent. And the mitochondrium tends to fill all the space available. Is also noticeable that the infected cells

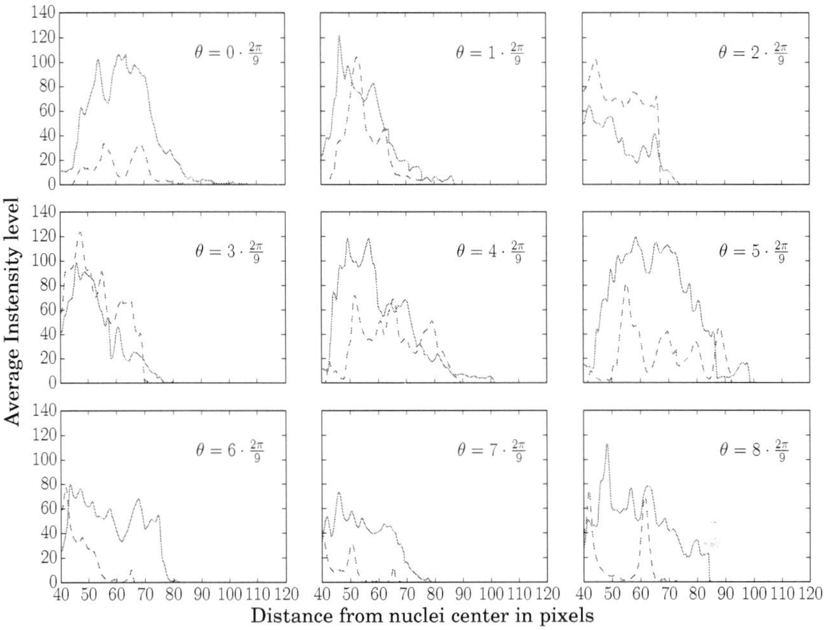

Figure 5. Average density of mitochondria $\rho(r, \theta, t)$, for $t = 0$ min. Mock cell mitochondrial density average is in blue (color online) dashed line and infected mitochondrial density is in red continuous line.

show more local maximum, which suggest that the distribution is somehow disorganized (more random).

These findings imply that a general structural change in mitochondrium distribution is caused by viral infection and it can be evidenced directly by examination of a cell's picture. The clustered behavior presented by mitochondrium on mock cells implies that they are grouped when normal function of the cell is in process. This is an organized distribution. On the other hand, the lack of clusters in infected cells shows that when infected the mitochondrium distribution became erratic, maybe random, which will be associated with a lack of organization.

A possible way to detect the presence of a viral infection will be to measure the level of randomness present in mitochondrium distribution. Remembering that $\rho(r, \theta, t)$ is experimentally measured through the pixel intensity in each video frame, it is found that the randomness in mitochondrium distribution will be the same than the randomness in pixel intensity distribution of the segmented image.

An image on gray scale is described digitally in terms of intensity values ranging from 0 (black) to 255 (white), a total of 256 possible shades of gray, each one of those possibilities is known as a level of intensity. A common descriptor used to classify a picture is the *entropy*

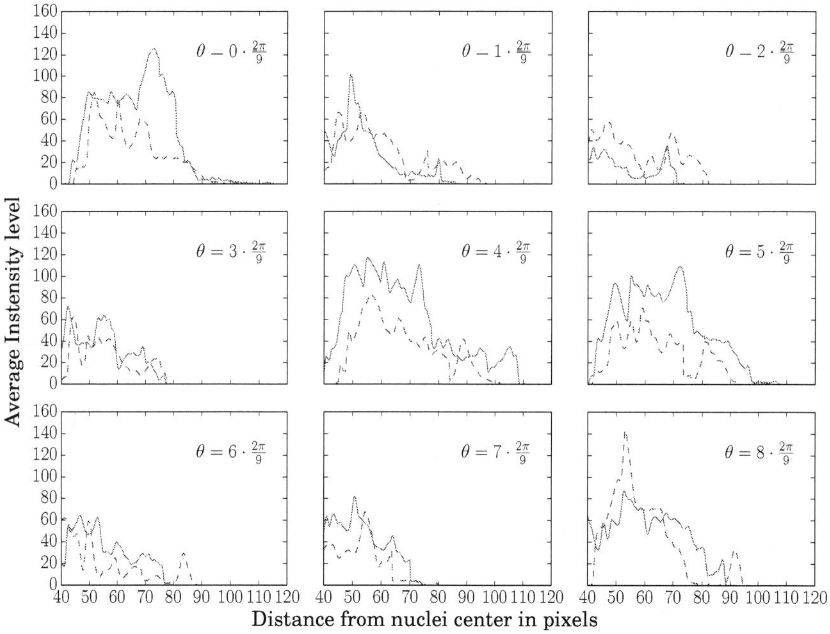

Figure 6. Average density of mitochondria $\rho(r, \theta, t)$, for $t = 320$ min. Mock cell mitochondrial density average is in blue (color online) dashed line and infected mitochondrial density is in red continuous line.

$$E = -\sum_{i=0}^{255} P(i)\log_2\left(P(i)\right). \tag{15}$$

This is a measure of how "random" the levels of intensity are distributed on a gray scale picture. $P(i)$is the probability of finding the i–th level of intensity inside the image. As mock cells shows more order than infected cells in $\rho(r, \theta, t)$, it can be expected that for the entropy in mock cell's image to be lesser than the entropy in the image of an infected one.

Another commonly used descriptor for images is *uniformity*, defined as:

$$u = \sum_{i=0}^{255} P(i)^2, \tag{16}$$

which is a measure of how much the levels of intensity change through the image. This descriptor is maximum if all the image presents one single level, and decreases with the level changes. By carefully looking at Figures 5, 6, and 7 the reader will note that infected cells shows more oscillations in the density of mitochondrium (local peaks). In terms of pixel intensity, it

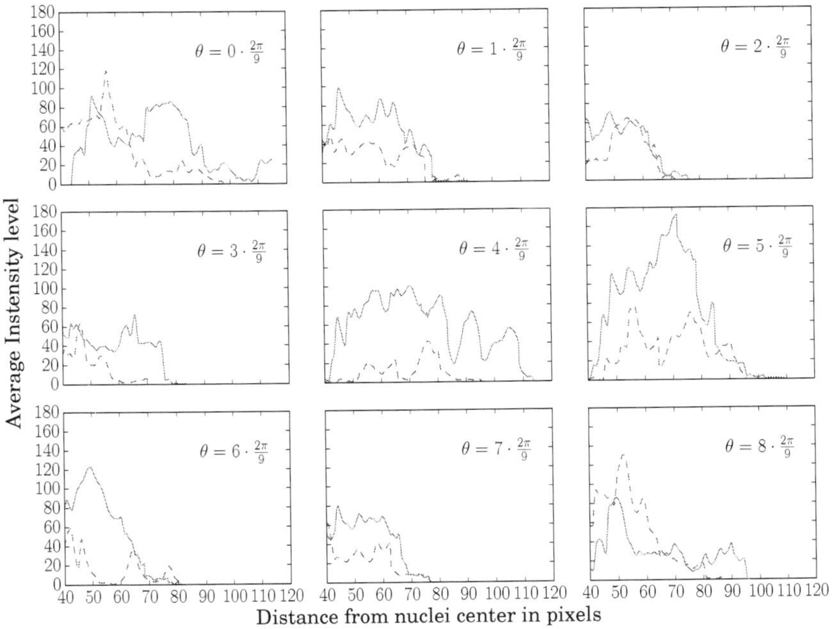

Figure 7. Average density of mitochondria $\rho(r, \theta, t)$, for t = 720 min. Mock cell mitochondrial density average is in blue (color online) dashed line and infected mitochondrial density is in red continuous line.

means that the intensity is changing more frequently, so the tone in the pictures is less uniform. Then it can be expected from the uniformity descriptor on the picture of an infected cell to be low.

In Figure 8, a plot on the uniformity vs entropy parameters space shows the computed values for those image descriptors for all studied cell in all frames. It can be seen that the mock and infected cells occupy mainly different regions on parameter space. So, these descriptors constitute a promising candidate to be a feature vector (or a part of it) in a machine learning code designed to classify infected cells.

5. Conclusion

A detailed analysis of the mitochondrium distribution around the nuclei for seven infected cells and four mock cells in nine videos has been performed. The study shows that mock cells clusters its mitochondrium and present an organized distribution in space. The organized character of the mitochondrium density distribution is maintained through time. At variance, infected cells loose these organized characteristics and the distribution of mitochondrium become erratic. This suggests that the mitochondrium are clustered when the healthy cell is

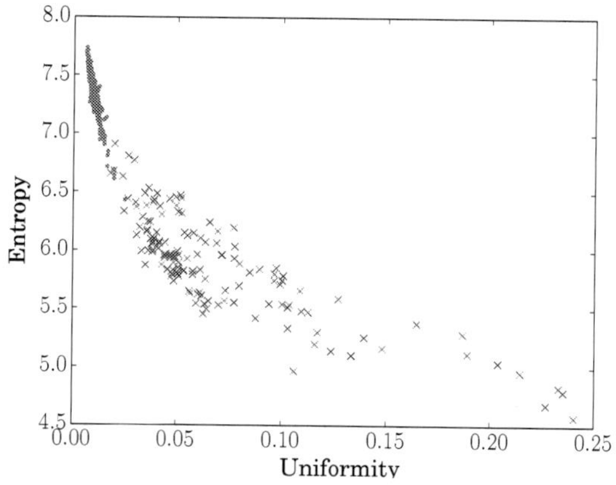

Figure 8. (Color online) Parameter space graph for Mock cell (blue x) and infected cells (red dots). Each marker represents the value for a single cell in a single time instant.

performing its natural process. But when a DENV infection is affecting the cell, those natural process are interrupted and it is reflected in the way how mitochondrium behaves. From this analysis, two image attributes are found to be suitable to be used as *features* in a ML classifier between infected and mock cells. These features are simple common image processing descriptors: *entropy* and *uniformity*, whose computation is easy and fast. Entropy is related with the randomness presented in the gray tones of the image and uniformity is related with the prevalence of a single gray tone. Both image attributes in a LCI photograph taken over a cell culture prepared with coloured cells are directly related with the mitochondrium density distribution behavior $\rho(r, \theta, t)$. The presence of clusters in mock cells and the softer $\rho(r, \theta, t)$ behavior are translated to higher values of uniformity and lower values of entropy image descriptors than those present in infected cells images.

Acknowledgements

This research was supported by COLCIENCIAS grant 111554531592 from the Colombian government. JCGG was the recipient of a Full-Time Professor Program (Exclusive Dedication) for the Medicine Faculty at University of Antioquia for 2014–2015.

Author details

Juan Carlos Cardona-Gomez[1*], Leandro Fabio Ariza-Jimenez[2] and
Juan Carlos Gallego-Gomez[2*]

*Address all correspondence to: juanc.gallegomez@gmail.com; jccg77@gmail.com

1 Optical Spectrometry Group, Faculty of Basic Sciences, Universidad del Atlántico, Barranquilla, Colombia

2 Translational and Molecular Medicine Group, Medellín Medical Research Institute, Faculty of Medicine, Universidad de Antioquia, Medellín, Colombia

References

[1] Brandenburg B., Zhuang X. Virus trafficking – learning from single-virus tracking. Nat Rev Microbiol. 2007;5 (3):197-208. DOI: 10.1038/nrmicro1615

[2] Lippincott-Schwartz J., Patterson G. H. Development and use of fluorescent protein markers in living cells. Science. 2003;300 (5616):87-91. DOI: 10.1126/science.1082520

[3] Frigault M.M., Lacoste J., Swift J.L., Brown C.M. Live-cell microscopy – tips and tools. J Cell Sci. 2009;122 (6):753-767. DOI: 10.1242/jcs.033837

[4] Wyckoff J., Gligorijevic B., Entenberg D., Segall J., Condeelis, J. High-resolution multiphoton imaging of tumors in vivo. In: Spector D.L., Goldman R.D., ed. Live Cell Imaging: A Laboratory Manual. 2nd edn. New York: CSHL Press, Cold Spring Harbor; 2010. p. 441-461.

[5] Lacey A.J., ed. Light microscopy in biology: A Practical Approach. 1st ed. IRL Press; 1989. 348 p.

[6] Duda R.O, Hart P.E., Stork D.G. Pattern Classification. 2nd ed. Wiley-Interscience; 2000. 680 p.

[7] Webb A.R. Statistical Pattern Recognition. 2nd ed. Wiley; 2003. 514 p.

[8] Bishop C. Pattern Recognition and Machine Learning. 1st ed. New York: Springer; 2007. 738 p.

[9] Mitchell T.M. Machine Learning. 1st ed. New York: McGraw-Hill; 1997. 414 p.

[10] Dorigo M., Schnepf U. Genetics-based machine learning and behavior-based robotics: a new synthesis. IEEE Trans Syst Man Cybernetics 1993;23 (1):141 154. DOI: 10.1109/21.214773

[11] Ball N.M., Brunner R.J. Data mining and machine learning in astronomy. Int J Mod Phys D. 2010;19 (07):1049-1106. DOI: 10.1142/S0218271810017160

[12] Clark M. Application of machine learning principles to modeling of nonlinear dynamical systems. Proc Arkansas Acad Sci. 1994;48:36-40.

[13] Neumann B., Held M., Liebel U., Erfle H., Rogers P., Pepperkok R., et al. High-throughput RNAi screening by time-lapse imaging of live human cells. Nat Meth. 2006;3 (5):385-390. DOI: 10.1038/nmeth876

[14] Jones T. R., Carpenter A. E., Lamprecht M. R., Moffat J., Silver S. J., Grenier J. K, et al. Scoring diverse cellular morphologies in image-based screens with iterative feedback and machine learning. Proc Nat Acad Sci. 2009;106 (6):1826-1831. DOI: 10.1073/pnas. 0808843106

[15] Carpenter A E, Jones T.R., Lamprecht M.R., Clarke C., Kang I.H., Friman O., et al. CellProfiler: image analysis software for identifying and quantifying cell phenotypes. Genome Biol. 2006;7 (10):R100. DOI: 10.1186/gb-2006-7-10-r100

[16] Tarca A L, Carey V.J., Chen X.-w., Romero R., Drăghici S.. Machine learning and its applications to biology. PLoS Comput Biology. 2007;3 (36):e116. DOI: 10.1371/journal.pcbi.0030116

[17] Haykin S. Neural networks: a comprehensive foundation. 2nd ed. Upper Saddle River, New Jersey: Prentice-Hall; 1999. 842 p.

[18] Mercer J. Functions of positive and negative type, and their connection with the theory of integral equations. Philosophical Transactions of the Royal Society of London. Series A, Containing Papers of a Mathematical or Physical Character. 1909;209:415-446.

[19] Rajaraman A, Ullman J.D. Mining of Massive Datasets. 1st ed. New York: Cambridge University Press; 2011. 326 p.

[20] Martínez-Gutierrez M., Castellanos J.E, Gallego-Gómez J.C. Statins reduce dengue virus production via decreased virion assembly. Intervirology. 2011;54 (4):202-216. DOI: 10.1159/000321892

[21] Padilla-S L., Rodríguez A., Gonzales M.M., Gallego-G, J C., Castaño-O, J.C. Inhibitory effects of curcumin on dengue virus type 2-infected cells in vitro. Arch Virol. 2014;159 (3):573-579. DOI: 10.1007/s00705-013-1849-6

[22] Orozco-García E., Trujillo-Correa A., Gallego-Gómez J. Cell biology of virus infection: The role of cytoskeletal dynamics integrity in the effectiveness of dengue virus infection. In: Najman S, ed. Cell Biology. 1st ed. InTech; 2015.

[23] Acevedo-Ospina H. Biología celular de la infección por virus dengue: Relación entre las mitocondrias y los virus [thesis]. Medellín, Colombia: Institute of Biology, University of Antioquia; 2014.

[24] Mahy B.W.J. The Dictionary of Virology. 4th ed. Academic Press; 2008. 520 p.

[25] Duda R.O., Hart P.E. Use of the Hough transformation to detect lines and curves in pictures. Communi ACM. 1972;15 (1):11-15. DOI: 10.1145/361237.361242

[26] Zhao T., Murphy, R.F. Automated learning of generative models for subcellular location: building blocks for systems biology. Cytometry Part A. 2007;71A (12):978--990. DOI: 10.1002/cyto.a.20487

[27] Wang Y.-l. Hahn K.M., Murphy R.F., Horwitz A.F. From imaging to understanding: frontiers in live cell imaging, Bethesda, MD, April 19–21, 2006. J Cell Biol. 2006;174 (4):481-484. DOI: 10.1083/jcb.200607097

[28] Tárnok A., Mittag A., Lenz, D. Clinical cytomics. In: Imaging, Manipulation, and Analysis of Biomolecules, Cells, and Tissues IV; February 21, 2006; SPIE Proceedings; 2006. p. 60880K-60880K-12. DOI: 10.1117/12.645024

[29] Kuznetsov A.V., Usson Y., Leverve X., Margreiter R. Subcellular heterogeneity of mitochondrial function and dysfunction: Evidence obtained by confocal imaging. Mole Cellular Biochemi. 2004;256-257 (1-2):359-365. DOI: 10.1023/B:MCBI.0000009881.01943.68

[30] Leonard A P, Cameron R.B., Speiser J L., Wolf B.J., Peterson Y.K., Schnellmann R.G., et al. Quantitative analysis of mitochondrial morphology and membrane potential in living cells using high-content imaging, machine learning, and morphological binning. Biochimi et Biophys Acta (BBA) - Mole Cell Res. 2015;1853 (2):348-360. DOI: 10.1016/j.bbamcr.2014.11.002

[31] Zhan M., Brooks C., Liu F., Sun L., Dong Z. Mitochondrial dynamics: regulatory mechanisms and emerging role in renal pathophysiology. Kidney Int. 2013;83 (4): 568-581. DOI: 10.1038/ki.2012.441

[32] Lihavainen E., Mäkelä J., Spelbrink J N, Ribeiro A S. Mytoe: automatic analysis of mitochondrial dynamics. Bioinformatics. 2012;28 (7):1050-1051. DOI: 10.1093/bioinformatics/bts073

[33] Reis Y., Bernardo-Faura M., Richter D., Wolf T., Brors B., Hamacher-Brady A., et al. Multi-parametric analysis and modeling of relationships between mitochondrial morphology and apoptosis. PLoS ONE. 2012;7 (1):e28694. DOI: 10.1371/journal.pone.0028694

[34] van Zutphen T., van der Klei I.J. Quantitative analysis of organelle abundance, morphology and dynamics. Curr Opin Biotechnol. 2011;22 (1):127-132. DOI: 10.1016/j.copbio.2010.10.015

Section 2

Molecular and Cellular Regulatory Mechanisms

Chapter 2

Epithelial Na^+,K^+-ATPase — A Sticky Pump

Jorge Alberto Lobato Álvarez, Teresa del Carmen López Murillo, Claudia Andrea Vilchis Nestor, María Luisa Roldán Gutierrez, Omar Páez Gómez and Liora Shoshani

Additional information is available at the end of the chapter

http://dx.doi.org/10.5772/61244

Abstract

Na^+,K^+-ATPase is an ATP-powered ion pump that establishes concentration gradients for Na^+ and K^+ ions across the plasma membrane in all animal cells by pumping Na^+ from the cytoplasm and K^+ from the extracellular medium. This heterodimeric enzyme, a member of P-type ATPases, is composed of a catalytic α-subunit with ten transmembrane domains and a heavily glycosylated auxiliary β-subunit. The Na^+,K^+-ATPase is specifically inhibited by cardiotonic steroids like ouabain, which bind to the enzyme's α-subunit from the extracellular side and thereby block the ion pumping cycle. Na^+,K^+-ATPAse generates ion gradients that establishes the driving force for the transepithelial transport of several solutes and nutrients. The effectiveness of this vectorial transport motivated by Na^+,K^+-ATPase depends on the integrity of epithelial junctions that are essential for the maintenance of the polarized localization of membrane transporters, including the lateral sodium pump. This chapter reviews the facts showing that, in addition to pumping ions, the Na^+,K^+-ATPase located at the cell borders functions as a cell adhesion molecule and discusses the role of the Na^+,K^+-ATPase β-subunit in establishing and maintaining cell–cell interactions. Furthermore, Na^+,K^+-ATPase is a multifunctional protein that, in addition to pumping ions asymmetrically and participating in cell–cell contacts, acts as specific receptor for the hormone ouabain and transduces extracellular signals. Thus, when bearing in mind with transporting epithelia phenotype, the importance of modulation of cell contacts by Na^+,K^+-ATPase can hardly be underestimated.

Keywords: Epithelial cells, Na^+, K^+-ATPase, Polarity, cell adhesion

1. Introduction

Epithelium is the name given to the cells that line a surface. Epithelia separate biological compartments with different composition, a fundamental role that depends on the establishment of occluding junctions. Thus, epithelial cells are always contiguous with one another and are usually joined by special junctions—the tight junctions. The functions of epithelia differ markedly. Although all form a barrier, some are much more impermeable than others. The trans-epithelial movement of ions and molecules is reached by the action of a multitude of specialized transporting proteins that are asymmetrically distributed on the apical or basolateral plasma membrane domains of epithelial cells. This remarkable polarity of epithelial cells depends on the selective insertion and the recycling of newly synthesized proteins and lipids into distinct plasma membrane domains and on the maintenance and modulation of these specialized domains once they are established during epithelial development. The two basic characters of epithelial transporting phenotype are polarity and tight junctions (TJs). Polarity offers the necessary direction for substances transported across epithelia to be absorbed or secreted. TJs guarantee that the transported substances do not leak back through the intercellular space [1, 2]. The Na^+,K^+-ATPase, known as the sodium pump, has been shown to play a central role in the transporting phenotype of epithelia. Today, we already know its crystal structure, the chemical composition as well as the spatial arrangement of its three subunits (α, β, and γ) (Fig. 1A), the relationship between ATP hydrolysis and ion movement, and several diseases related to its malfunction. Surprisingly, despite such detailed information, we keep finding new remarkable properties and new physiological functions of the Na^+,K^+-ATPase. The present chapter is focused on three recently found properties: its expression at the lateral membrane of epithelial cells due to the self-adhesive property of the β-subunit, its role as hormone receptor, and its ability to modulate several types of cell contacts.

2. Na^+,K^+-ATPase in epithelia

The history of Na^+,K^+-ATPase can be compared to a double-sided step ladder, one climbed by biologists that investigated its intrinsic mechanisms and subunits the other by accomplished physicists who first criticized but finally solved immense theoretical obstacles in the road toward active transport. Thus, for a long while, the peculiar composition of the cytoplasm was attributed to presumed membrane impermeability to Na^+. This alternative was invalidated right after the Second World War, when radioisotopes became available for biological research, and it was discovered that tracer Na^+ added to the bathing solution readily penetrates and distributes into the cell. This revived the question why Na^+ in the cytoplasm remains at a concentration much lower (~15 mM) than in the extracellular water (~140 mM). The simplest assumption was that there should be an enzyme in the plasma membrane that dissipates metabolic free energy to pick up Na^+ from the cell water and pump it out. This response was not readily accepted because it appeared in violation of the Curie principle: "Processes of different tensorial order cannot be coupled." In simple worlds, metabolism and ATP hydrolysis are chemical reactions, which at that time were taken to be *scalar* processes; hence, they could not drive a *vectorial* process like the extrusion of Na^+ from the cytoplasm. Later on, it was argued that, in fact, chemical reactions are vectorial at microscopic level. However, this vectoriality is masked at macroscopic scale, in particular, when working with homogenized tissues where pumps point in all directions. However, if they were ordered in a membrane, the asymmetry would be recovered, and a macroscopic flux would take place. A model put forward by Koefoed-Johnson and Ussing [3] proposed a pump located asymmetrically on the basal side of the epithelial cell that together with the specific Na-permeability of the outer cell membrane and the specific K-permeability of the inner facing membrane is responsible for the net movement of Na^+. This made theoreticians happy, yet where was "the pump," i.e., the membrane molecule, that would align and be responsible for the sided, asymmetrical movement of Na^+? On 1957, Jens Christian Skou prepared an extract of crab tissue that contained an enzyme that splits molecules of ATP (hence deserving the name "ATPase") into ADP + Pi, provided the medium contains K^+ and Na^+ ions at concentrations that compare with those in the cell and in the surrounding extracellular space. Therefore, the enzyme was aptly named Na^+,K^+-ATPase. Interestingly, Skou [6] was able to inhibit the ATP splitting activity of his extract by adding *ouabain*, a substance of vegetal origin that was found a few years earlier to inhibit active potassium and sodium transport in erythrocyte membrane [5]. By performing the Na^+/K^+ translocations cyclically, Na^+,K^+-ATPase transfers those ions in a net amount toward the extracellular medium and toward the cytoplasm, respectively, so it was justified to call it "pump" [6]. Another experimental/theoretical conflict occurred with sugars and amino acids transported in a net amount across epithelia. Since this transport occurs in a net amount and can be inhibited with ouabain, for a while, it was taken as a proof that there exists a glucose pump as well as other pumps for diverse amino acid species. Yet eventually, it was demonstrated that carriers for sugars and for amino acids are not pumps as they are not *directly* coupled to metabolism. Therefore, the Na^+,K^+-ATPase is the *primum movens*, responsible for the exchange of substances between the metazoan and the environment, across transporting epithelia, as well as for net exchange between the internal milieu and the cytoplasm.

2.1. Structure–function relationship of Na^+,K^+-ATPase

Na^+,K^+-ATPase is expressed in all animal cells and is one of the most important members of the P-type ATPases. The Na^+,K^+-ATPase creates the Na^+ and K^+ concentration gradients across the plasma membranes of most higher eukaryotic cells. Per cycle, it pumps three Na^+ ions out and two K^+ ions into the cell, coupling the energy derived from the hydrolysis of one ATP molecule. The Na^+ and K^+ gradients originated and maintained by Na^+,K^+-ATPase are the energy source for secondary active transport, which are used for the maintenance of cell osmolarity and volume, for the generation of action potentials along nerve cells, and for many other cellular purposes. The functional Na^+,K^+-ATPase is a heterodimer of α- and β-subunits. In addition to the $\alpha\beta$ heterodimer, there are tissue-specific regulatory γ-subunits, also known as FXYD proteins [7]. In this section, we will review the general structural and functional characteristic of each subunit and how the pumping of Na^+ and K^+ is achieved. Important findings about the function are briefly discussed in the light of several X-ray crystallography studies of Na^+,K^+-ATPase published in recent years [8–10] (Fig. 1A).

2.1.1. The α-subunit

The catalytic α-subunit is composed of approximately 1000 amino acid residues with a molecular mass of about 110 kDa. Since the first sequencing and cloning of the α-subunit of the Na^+,K^+-ATPase [11], many biochemical studies pointed out to a model of ten transmembrane domains (M1–M10) and three cytoplasmic domains: A (actuator), N (nucleotide binding), and P (phosphorylation domain) (Fig. 1A). Another myriad of studies identified motifs crucial for cation binding and conformational transitions [12]. Four distinct isoforms of the α-subunit have been identified (α_1–α_4) in which sequence differences are minor. Each isoform has different kinetic properties which may be essential in adapting cell Na^+,K^+-ATPase activity to specific physiological requirements [3]. The major form α_1 is found in most tissues and is the main or only form in kidney and most other epithelia.

The Na^+,K^+-ATPase *A domain* consists of the N-terminal segment plus the loop between M2 and M3 (Lys212-Glu319), which form a distorted jelly roll structure plus two short helices in the 40 residues of the N-terminal segment. The movements of this domain, especially of the loop M2–M3, are determinants for the TM conformational rearrangements, needed for the occlusion and release of cations (see Fig. 1B) [14]. *N domain* contains the ATP-binding site and extends from the phosphorylation site Asp376, with the 377-KTGTL sequence, to the C-terminal 593-DPPR hinge motif. Finally the *P domain* can be described as a six-stranded parallel β-sheet and contains three important motifs: 376-DKTGTL, which contains the aspartic acid residue that phosphorylates during catalysis; 617-TGD on strand 3, which associates with the phosphorylation motif during the conformational transition E1 to E2; and 715-TGDGVND, which terminates the sixth β-strand of the P-domain where Asp717 is required for binding of Mg^{2+} and phosphorylation (Fig. 1B).

The first crystal structure of Na^+,K^+-ATPase α-subunit published was that of Rb^+-bound pig renal Na^+,K^+-ATPase. Each of the three cytoplasmic domains and transmembrane helices of α-subunit are superimposable with Ca^{2+}-ATPase (SERCA). The two sites for K^+ binding are found between helices M4, M5, and M6, and many of the residues involved had already been identified in various studies as cation-binding residues [8]. The first complete high-resolution crystal structure of Na^+,K^+-ATPase was obtained in a potassium-bound state and provided further detail into the molecular basis for K^+ specificity [9] (Fig. 1A). The carboxy terminal part of the L7/8 loop is the primary interaction site with the β-subunit, in which a consensus sequence 901 SFGQ, proposed as a key interaction site is located [15].

2.1.2. *The β-subunit*

The β-subunit of the Na^+,K^+-ATPase was initially identified as a glycoprotein associated with the α-subunit in purified functional enzyme preparations [16]. The association between α- and β-subunits is relatively strong and remains stable in most non-ionic detergents. All of the known β-subunit species and isoforms share a common domain structure: a short N-terminal cytoplasmic tail, a single transmembrane segment, and a large extracellular C-terminal domain containing six extracellular cysteine residues forming three disulfide bridges, whose locations are completely conserved among the isoforms. Na^+,K^+-ATPase β-subunit is composed of approximately 310 residues with an apparent molecular mass of 55 kDa due to N-glycosylation. The β-subunit of the Na^+,K^+-ATPase has three isoforms designated β_1, β_2, and β_3. Among these isoforms, there are various degrees of difference, and although all β-subunits are glycosylated, the number of N-glycosylation sites varies with the isoform. The β_1 isoform of

http://dx.doi.org/10.5772/61244

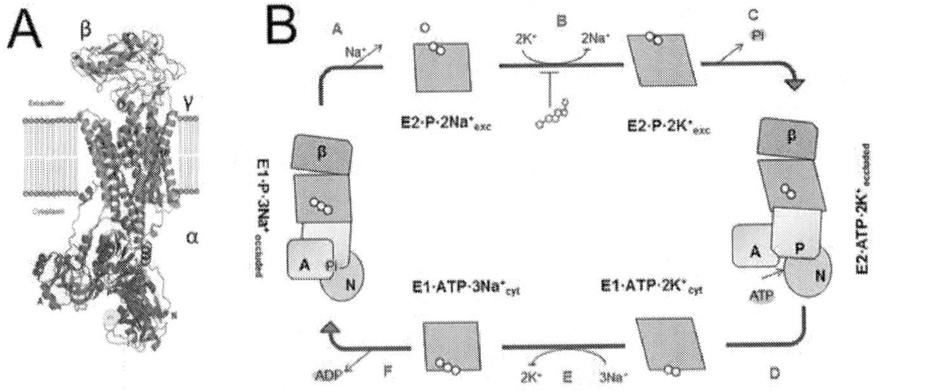

Figure 1. Structure of the Na^+,K^+-ATPase. (A) Ribbon model of the crystal structure of shark Na^+,K^+-ATPase [9; PDB code: 2ZXE], indicating α-subunit transmembrane domains in gray and cytoplasmic domains A, N, and P in red, green, and blue, respectively. The β-subunit is colored in orange and lacks most of its N-terminal cytoplasmic domain as in the crystal. The γ-subunit is colored in magenta. Alpha subunit transmembrane helices are numbered with the exception of TM 1 and 9, which are not visible from this projection. The two potassium ions occluded in the crystal structure are depicted in yellow. In the N domain, the ATP binding site is also indicated in yellow. (B) Schematic depiction of the catalytic cycle of Na^+,K^+-ATPase. A structural rearrangement, especially in domain A (pink-colored), is suggested by the cartoons at both conformational states E1 and E2. For simplicity, the rest of the cycle stages are represented only with the TM region in gray. (A) The outward transport of three Na^+ ions is coupled to the E1 to E2 transition. (B) Two K^+ ions bind at binding sites positioned to the extra cellular space. (C) Extracellularly bound K^+ ions activate dephosphorylation that in turn results in ion occlusion. Ouabain (cardiotonic steroid silhouette in brown) binds at this conformational state halting the cycle. (D) ATP with a low affinity triggers the acceleration of inward transport of K^+ ions through the pore as the pump enters the E1 conformational transition with low affinity for K^+. (E) Three Na^+ ions bind the intracellularly oriented sites. (F) Phosphorylation from ATP occurs, and Na^+ ions are occluded. Several transitional substates exist during the cycle; nevertheless, they are not depicted for simplification.

Na^+,K^+-ATPase, which is consistently predicted to have three N-linked glycosylation sites, has been most extensively studied. All three consensus sites of β_1 are glycosylated [17, 18], and the oligosaccharides are terminally sialylated. Treuheit et al. [19] showed by mass spectrometry that the oligosaccharides of the β_1-subunit from dog and lamb kidney are of tetra-antennary structure with extensions of 2–4 *N*-acetyllactosamine units, and the units of extension seemed to differ between the dog and the lamb β_1-subunits. Nevertheless, detailed information is not available for the oligosaccharide composition of the two other isoforms, but it seems clear that there is a high degree of species variability for the β_2 isoform too. This isoform contains 7–9 N-glycosylation sites. It holds the high mannose-type carbohydrate epitopes L3 and L4 [20] also present on the neural recognition molecules L1, MAG, and P0 that mediate adhesion among neural cells [21]. Indeed, the β_2 subunit was originally identified as an adhesion molecule, AMOG, in glial cells [22].

The essential function of the β-subunit is acting as the molecular chaperone of the α-subunit. Na^+,K^+-ATPase α-subunit is inactive without its β-subunit. It has been demonstrated that the association of the β-subunit facilitates the correct packing and membrane integration of the newly synthesized α-subunit. Also, as the intimate partner of α-subunit, it modulates cation-binding affinity [23, 24]. The complete crystal structure of Na^+,K^+-ATPase explains, at least partially, previous implications of the β-subunit in modulation of cation transport. The transmembrane helix of the β-subunit runs slightly separated from those of the α-subunit and is rather inclined than perpendicular to the membrane. It forms several interactions with M7 and M10 helices of the α-subunit. The extracellular domain of the β-subunit contains the well-conserved YYPYY motif that mediates several salt bridges with α-subunit L7/8 loop and also contains at least three additional clusters of residues interacting with α-subunit [9]. In epithelia, in addition to the classical chaperone function of the β-subunit, a cell-to-cell adhesion function has been ascribed to the β_1 subunit. This novel role of the β_1-subunit will be described in a separate part of this chapter.

2.1.3. The γ-subunit

In addition to the αβ heterodimer, there are tissue-specific regulatory γ-subunits, which are small membrane proteins characterized by an FXYD sequence, and therefore known also as FXYD proteins, of approximately 80–160 residues [24]. FXYD proteins mainly consist of a single transmembrane helix and an N-terminal extracellular domain where the FXYD motif is located. This domain anchors to the β-subunit extracellular and transmembrane domains. These proteins modulate the function of Na^+,K^+-ATPase adapting kinetic properties of cation active transport to the specific needs of different tissues [7]. The most studied FXYD proteins are FXYD1 or phospholemman mainly expressed in heart and skeletal muscle and is involved in heart contractility. In epithelia, kidney-specific FXYD2 decreases affinity of Na^+,K^+-ATPase for sodium and FXYD4 or CHIF, expressed in colon and kidney epithelia acts as a modulating several ion transport mechanisms that have Na^+,K^+-ATPase as a common denominator [26, 27].

2.2. The pumping catalytic cycle

Ion movements through the Na^+,K^+-ATPase have been studied by biophysical experiments for many years [28–30]. Those studies were incorporated into the conceptual framework called post-Albers cycle that depicts the sequence of reaction steps that couple ion transport and ATP hydrolysis (Fig. 1B). Na^+ and K^+ transport follow a "ping-pong" mechanism, wherein the two ion species are transported sequentially. Pumping of ions is achieved by alternation between two major conformational states, E1 and E2 [31, 32]. In E1, the cation-binding sites have high affinity for Na^+ and face the cytoplasm; in E2, the cation-binding sites have low affinity for Na $^+$ but high affinity for K^+ and face the extracellular. As it is the case for all P-type ATPases, Na $^+,K^+$-ATPase autophosphorylates and dephosphorylates during each reaction cycle. In E1, after three Na^+ ions are bound at the cytoplasmic face, the phosphoryl group is transferred to a conserved aspartic residue in the P-domain. At this point, the pump enters the E1P state with occluded Na^+ ions, and when ADP leaves, another conformational change occurs, and the Na^+ ions are released at the extracellular face. At this stage, the enzyme is no longer sensitive to ADP addition but to aqueous hydrolysis (E2P state) so that Pi is released at the catalytic site and so are two K^+ ions occluded at the extracellular face. To complete the cycle, ATP binds the phosphorylation site leading to the departure of the two K^+ ions in the cytoplasm. At this point, the pump returns to the E1 state with high affinity for Na^+ ready to launch a new catalytic cycle (Fig. 1B).

The first crystal structures of Na^+,K^+-ATPase have been obtained in the E2 state, which is more stable [8, 9]. Nevertheless, they lack information about the Na^+-bound state and in particular the location of the third Na^+ site. Two recently published crystals both of Na^+,K^+-ATPase from pig kidney are of E1 states. These crystals are stable analogues of the transition state (E1P-ADP $^+$▪3Na^+) preceding E1P▪3Na^+. The molecular comparison of the two states, E1 and E2, show that the α-subunit suffers important conformational changes in its cytoplasmic domains, especially in the A domain, which is rotated around an axis nearly perpendicular to the membrane. The transmembrane helices involved in cation binding also undergo important conformation changes, most of all TM 4, 5, and 6 (Fig. 1B). Based on these structural evidences, the third Na $^+$-binding site has been clearly localized, and a cooperative process for the sequential binding of Na^+ has now been formulated in detail [33, 34].

2.3. Mechanism of Na^+,K^+-ATPase polarity in epithelia

The apical and basolateral plasma membrane proteins of epithelial cells are synthesized in the endoplasmic reticulum (ER) and then sorted in the tans-Golgi network (TGN) to be sent into different carrier vesicles to apical or basolateral domain [35, 36]. The polarity of those routes depends significantly on specific signals encoded inside the membrane proteins. The basolateral proteins have short peptides sequences in the cytoplasmic domain. Some signals resemble endocytic signals (dileucine, YXXϕ, and NPXY), while others are unrelated to endocytic signals (the tyrosine motifs in LDL receptor [37] and the G-protein of the VSV [38]). Early studies demonstrated that the Na^+,K^+-ATPase, comprised of α- and β-subunits, is sorted in the TGN and delivered directly to the basolateral membrane without significant appearance at the apical surface in certain strains of the Madin–Darby canine kidney cells (MDCK) [39, 40]. Therefore, a basolateral signal was assumed to exist in the α-subunit of the Na^+,K^+-ATPase. The Na^+,K^+-ATPase and the H^+,K^+-ATPase are highly homologous ion pumps, yet in LLC-PK1 cells they are polarized to the basolateral and the apical domains, respectively. The polarized expression of chimeric constructs of the α-subunit of the H^+,K^+-ATPase and the Na^+,K^+-ATPase in LLC-PK1 cells has been studied [41–43]. Uncommonly, an apical sorting information in the α-subunit of the H^+,K^+-ATPase was recognized within the fourth transmembrane domain. Swapping this domain into the Na^+,K^+-ATPase resulted in the redirection of that basolateral pump to the apical surface of LLC-PK1 cells [44]. Nevertheless, these studies do not clarify whether the α-subunit of the Na^+,K^+-ATPase contains a basolateral sorting signal in its fourth transmembrane domain. Therefore, it seems that a non-canonical polarity signal is involved in the basolateral targeting of Na^+,K^+-ATPase.

Clathrin plays a fundamental role in basolateral sorting. It interacts with endocytic or basolateral proteins through a variety of clathrin adaptors [45]. It has been shown that the adaptor involved in basolateral protein sorting is the epithelial cell-specific AP-1B (adaptor protein 1B). Nevertheless, the basolateral localization of the Na^+,K^+-ATPase is independent of AP-1B expression because its localization was not significantly affected by knocking down clathrin expression and it remained localized to the basolateral surface in both the μ1B-deficient cell line LLC-PK1 [46] and in MDCK cells in which μ1B expression had been suppressed via RNAi [47]. By taking advantage of the SNAP tag system to reveal the trafficking itinerary of the newly synthesized Na^+,K^+-ATPase, it was shown that the basolateral delivery of the Na^+,K^+-ATPase is very fast (at 5 minutes after Golgi release, 50% of newly synthesized Na pump is colocalizing with the PM) and does not involve passage through recycling endosomes en route to the plasma membrane. Moreover, Na^+,K^+-ATPase trafficking is not regulated by the same small GTPases as other basolateral proteins [48]. Some membrane proteins may achieve polarity by selective retention at the apical or basolateral surface. Although less well understood, this polarity may reflect interactions with extracellular ligands or with intracellular scaffolds, such as cytoskeletal elements or arrays of PDZ domain-containing proteins [35, 49–51]. As described and discussed below, this is also the case of the epithelial Na^+,K^+-ATPase, which is retained at the lateral membrane domain due to *trans* adhesion of its β_1 subunits on neighboring cells.

3. Na^+,K^+-ATPase β subunit as an adhesion molecule

3.1. The β_1-isoform is a self-adhesion molecule in epithelia

The β-subunit is a glycoprotein of 40–60 kDa that was shown to be involved in the structural and functional maturation of the holoenzyme [52, 53] and subsequent transport of the α-subunit to the plasma membrane [54–56]. Ion transport requires the participation of both α- and β-subunits [54, 57]. The β-subunit has a short cytoplasmic tail, a single transmembrane segment, and a long extracellular domain heavily glycosylated, a typical structure of a cell-attachment protein [25]. Fig. 1A depicts the position and arrangement of the three subunits of Na^+,K^+-ATPase: α-subunit, β-subunit, and γ-subunit obtained by crystallography. Note that the β-subunit is mostly exposed toward the intercellular space, while most of the α-subunit is contained in the cytoplasm [9]. Observations made in MDCK cells suggested that the β-subunit is a cell–cell attachment protein: (1) As most transporting epithelia, the monolayer of MDCK expresses Na^+,K^+-ATPase polarized toward the basolateral side [58]. Nevertheless, confocal immunofluorescence analysis of Na^+,K^+-ATPase localization shows that the pump is not located on the *basal* domain of the plasma membrane, but only in the *lateral*, at cell–cell contacts (Fig. 2A). (2) Upon previous treatment with EGTA, the confocal images show the apparent single green line splits into two indicating that in order to express Na^+,K^+-ATPase at a cell–cell contact both neighboring cells have to contribute part of the enzyme. (3) The expression of the Na^+,K^+-ATPase at a given lateral borders is observed when both contributing neighboring cells are homotypic and from the same species, for instance, MDCK/MDCK (dog/dog, Fig. 2A) but not MDCK/NRK (dog/rat) [59] (Fig. 2B). (4) When CHO cells (fibroblasts from Chinese Hamster Ovary) were transfected with a gene coding for the β_1-subunit of the dog (CHO-dog β_1), these cells become more adhesive, as estimated by aggregation assays [60]. (5) On the other hand, mixed monolayers of MDCK and NRK dog β_1 show that MDCK cells expose the Na^+,K^+-ATPase at the heterotypic border (Fig. 2C).

All together, these observations indicated that the lateral localization of the Na^+,K^+-ATPase in MDCK cells depends on the recognition and adhesion between the β_1-subunits of neighboring cells [60] (Fig. 2D). Of course, the first question that arises is whether two corresponding β-subunits from different cells would get close enough to be able to span the intercellular space and interact directly as proposed. To answer this question, several protein–protein interaction

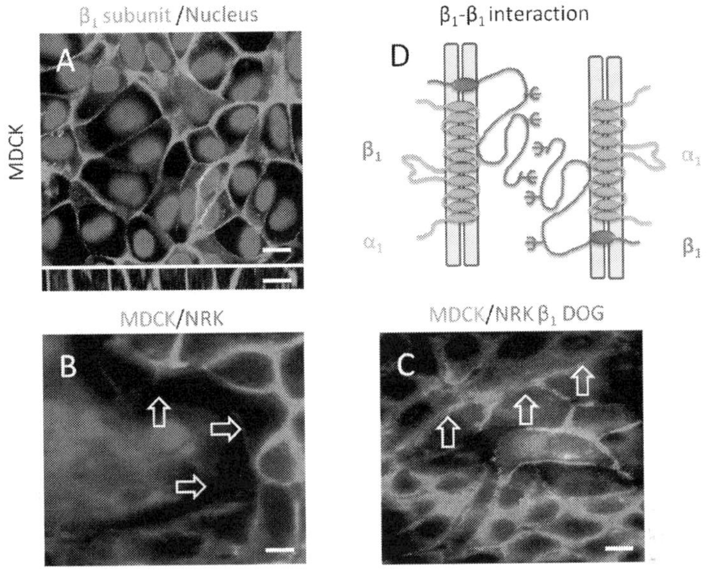

Figure 2. Hints to propose a model for the polarized distribution of Na^+,K^+-ATPase in transporting epithelia. (A) Monolayer of MDCK cells in a horizontal and a transversal section. Na^+,K^+-ATPase is stained in green, and nuclei in red, showing that the pump is expressed on the lateral membrane of the cells. (B) A confocal image of a monolayer prepared with a mixture of MDCK cells and NRK (normal rat kidney) cells; notice that the MDCK cells surrounding the NRK cell (previously stained in red with CMTMR) in the center only express their Na^+,K^+-ATPase on the membrane contacting MDCK cells, but not on the side contacting the NRK epithelial cell (indicated by arrows). (C) A confocal image showing a mixture of MDCK and NRK cells transfected with dog β_1-subunit. Arrows indicate the presence of Na^+,K^+-ATPase at heterotypic borders. (D) Proposed model for the polarized distribution of Na^+,K^+-ATPase in transporting epithelia. Scheme showing Na^+,K^+-ATPase α- and β-subunits expressed at the lateral border, where they are anchored by the β-subunits interaction at the intercellular space. Scale bar: 10 μm.

assays have been performed: (1) By pull-down assay, it was shown that dog β_1-subunit immobilized on Ni-beads could specifically bind to the soluble extracellular domain of β_1-subunits of the same animal species (dog). (2) Co-IP experiments have shown that rat β_1-subunits on NRK cells co-precipitate with rat YFP-β_1 subunit transfected in MDCK cells. (3) FRET (fluorescence resonance energy transfer) analysis of monolayers with a mixed population of MDCK cells transfected with a β_1-subunit fused to a cyan fluorescent protein (*CFP*), or with a β_1-subunit fused to yellow fluorescent protein (YFP), has shown that energy can be transferred from the first to the second cell type; in other words, two β_1-subunits can interact directly at <10 nm, thereby anchoring the whole enzyme at the cell membrane facing the intercellular space. Taken together, these evidences [61] supported by works from other groups [62-65] indicated that the β_1 subunit is indeed an adhesion molecule in epithelia.

In Moloney sarcoma virus-transformed MDCK cells (MSV-MDCK) that have an invasive phenotype, the level of Na^+,K^+-ATPase β_1-subunit is reduced as well as the expression level of E-cadherin. As expected, these transformed cells are also deficient in tight and desmosome junctions. Interestingly, transfection of both E-cadherin and Na^+,K^+-ATPase β_1-subunit induces the formation of junction complexes, reestablishes epithelial polarity, and suppresses invasiveness and motility, suggesting that β-subunit and E-cadherin are required to maintain the polarized epithelial phenotype [66]. Furthermore, stable adherens junctions are a requisite for proper tight junction function. In this regard, improving the Na^+,K^+-ATPase β_1–β_1 interaction by reducing the complexity of the N-glycans of the β-subunit increases the resistance to detergent extraction of junction proteins and decreases the paracellular permeability. In other words, the fewer the branches are in β-subunit's N-glycans, the tighter are the intercellular junctions. Conversely, the impairment of the β_1–β_1 binding by removing the N-glycans or altering the amino acid sequence of one of the interacting proteins decreases detergent resistance and increases the paracellular permeability, indicating that stability of adherens, and in turn tight junctions, does depend on β_1–β_1 interaction [65, 67].

Studies in Drosophila have also shown that the β-subunits (in drosophila are named Nrv1, Nrv2, and Nrv3) are determinant of the Na^+,K^+-ATPase subcellular localization as well as function. Of the three Drosophila isoforms, Nrv1 and Nrv2 are localized in epithelia, while Nrv3 is expressed in the nervous system. Remarkably, while Nrv1 is expressed in the basolateral membrane of almost all epithelial cells, Nrv2 is localized at the septate junctions (tight junctions in insect) and co-localizes with coracle [68]. Furthermore, it has been shown that the extracellular domain of Nrv2 regulates the function of septate junctions and the size of the tracheal tube in a free manner independent of the pumping task [69].

3.2. β_2/AMOG is a heterophilic adhesion molecule in nervous system

The Na^+,K^+-ATPase β_2-subunit was first described in the nervous system. Schachner's group identified a cell surface glycoprotein and named it as Ca^{2+}-independent adhesion molecule on glia (AMOG). AMOG was shown to mediate the neuron-to-astrocyte adhesion in the process of granule cell migration [70–72]. Further analysis revealed that AMOG is an isoform of the Na^+,K^+-ATPase β-subunit, named as the β_2-subunit [22]. A remarkable characteristic of the β_2-subunit is the multiple N-glycosylation sites in the extracellular domain [13]. Treatment with endoglycosidase H produces the shift of the apparent molecular weight from 50 to 35 kDa [70,20]. In a mass spectrometry analysis of the endoglycosidase H, released oligosaccharides from the β_2-subunit three molecular ions were found corresponding to oligosaccharides composed of one N-acetylglucosamine and 5, 6, or 7 mannoses [20]. The β_2-subunit promotes the neurite outgrowth by AMOG-to-neuron binding [73]. Schachner's group has assayed different partners for AMOG association in *trans*. They found that AMOG-containing liposomes only bind to small cerebellar neurons. When L1 and N-CAM antibodies were added to a monolayer of cerebellar neurons, none of these antibodies inhibited binding of AMOG-containing liposomes to neurons. Also, cells preincubation with an AMOG-antibody prior to addition of AMOG-containing liposomes did not reduce AMOG-containing liposomes-to-neurons adhesion [71]. These experiments suggest that neither L1 nor N-CAM is the β_2-subunit ligand; thus, AMOG/β_2-subunit is a heterophilic CAM. As mentioned above, β_1–β_1 adhesion in epithelial cells is homophilic. Accordingly, when assaying β_1 to β_2 adhesion, we found a null binding between these two isoforms, as well as between two β_2-subunits (our unpublished results). These findings are in agreement with previous studies [71]. As heterophilic CAM, the β_2-subunit was shown to *cis*-interact with an oligomannose binding lectin, basigin. Basigin or CD147 is an ancillary protein of the monocarboxylate transporters 1, 3, and 4-isoforms [74–77] and, as a receptor molecule for high mannose carbohydrates, basigin binds specifically with oligomannoside carrying glycoproteins and neoglycolipids [78]. Kleene and coworkers showed that PrP, the AMPA receptor subunit GluR2, the astroglial α_2/β_2 ATPase, basigin, and the MCT1 form a functional complex at the plasma membrane of astrocytes. In this regard, the β_2-subunit and basigin interact by means of the carbohydrate structure of the β_2-subunit. The functional interplay of PrP, GluR2, the α_2/β_2 ATPase and basigin regulates the lactate transport via MCT1. Moreover, they observed that disturbing the oligomannose-mediated interaction of the β_2-subunit and basigin leads to a deregulated and thus elevated glutamate-independent lactate transport [79].

3.3. β_2 isoform and apical polarity

The α_1-subunit holds a basolateral sorting signal that commands the traffic of the epithelial sodium pump to this membrane domain [43]. However, the role of other α-subunit isoforms in the sorting of the Na^+,K^+-ATPase has not been studied yet. On the other hand, little is known about sorting signals in any of the β-subunit isoforms, yet the β_1 and β_3 isoforms have exclusive basolateral localization in epithelial cells [80], and instead, the apical distribution of the sodium pump correlates with the expression of the β_2 isoform [81–84]. In this regard, studies from our laboratory provide evidence showing that the apical polarity of the Na^+,K^+-ATPase in the retinal pigment epithelium (RPE) is related to the expression of the α_2- and β_2-subunits (Fig. 3A and B). Moreover, the time-dependent β_2-subunit expression in the RPE model cells ARPE-19 correlates with the epithelialization of these cells (our unpublished results).

As we mentioned before, the β_2-subunit possess up to 9 N-glycosylation sites (upon the species). Numerous studies have indicated the role of N-glycans in the polarity mechanism of apical proteins. For instance, the mutagenic removal of N-glycosylation sites in the gastric H^+,K^+-ATPase β-subunit [85], bile salt export pump [86], and glycine transporter 2 [87, 88], significantly decreased their apical content and increased their intracellular accumulation. Also, it has been shown that addition of N-glycans to various proteins changed their cellular localization toward the apical membrane domain. For example, a truncated occludin and a chimeric ERGIC-53 residing inside the Golgi in their nonglycosylated forms were apical redistributed after addition of N-glycans [89]. Indeed, engineering the β_1-subunit by adding the N-glycosylation sites of the β_2 isoform leads to apical localization of the pump in HGT-1 cells [90]. All these evidences are consistent with the important role of N-glycosylation in apical polarization of Na^+,K^+-ATPase.

3.4. Structural insights into the self adhesion mechanism of Na^+,K^+-ATPase β_1 subunits

The shark Na^+/K^+-ATPase crystal structure in the E2 state published by Shinoda and coworkers was the first resolving the atomic structure of the extracellular domain of the β-subunit (PDB: 2ZXE) [9]. The extracellular C-terminal domain of the protein folds into an Ig-like β-sheet sandwich as predicted in silico [91]; actually, deletion of this C-terminal domain abolishes the

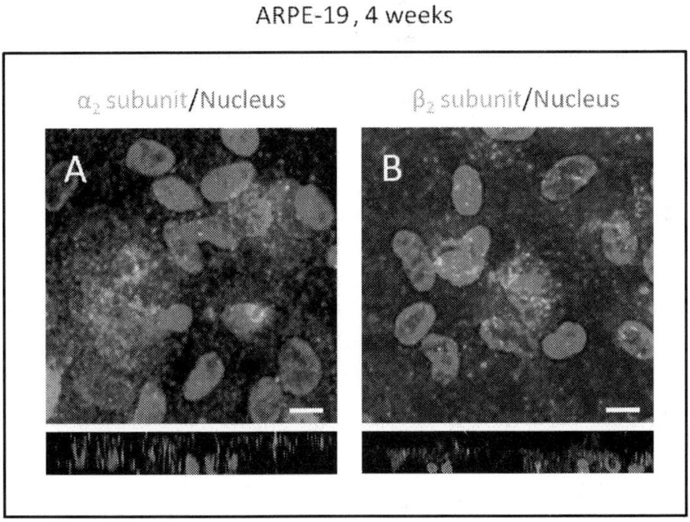

Figure 3. Na^+,K^+-ATPase expression at the apical domain of ARPE-19 cells (human retinal pigment epithelium). ARPE-19 cells were cultured on laminin-coated inserts for 4 weeks and treated for IF analysis. Confocal image of a monolayer stained with specific antibody against the α_2 subunit (A) and against β_2 subunit (B). Notice the preferential distribution on apical domain of both subunits. Scale bar: 10 μm

β_1 adhesion capacity (unpublished observations). However, a large number of adhesion and nonadhesion proteins contain domains with an immunoglobulin-like topology (CATH database). Structural alignments of the β_1 subunit extracellular domain against other well-studied cell adhesion molecules reveal no structural homologue of β-subunits of any kind. Detailed inspection of the ectodomain structure uncovers several features distinctive to β-subunit family members. Namely, its Ig-like fold has a unique topology given that its β-sheet sandwich is interrupted by a long α-helix secondary structure and has an atypical β-sheet disposition in relation to classical Ig folds. Also, the β-subunit fold contains extensive loops and therefore its length is twice as that of a typical Ig domain. Furthermore, the β_1 subunit is structurally compromised with the catalytic α-subunit in such a way that the C-terminal fold must be more rigid than the typical flexibility of whole adhesion domains such as in cadherins. Altogether, these observations suggest that the β-subunit of the Na^+,K^+-ATPase must possess an adhesion mechanism that is particular to this family, as shown in Figure 4.

The first attempt to clarify this adhesion mechanism on a molecular base is related to the regions of the ectodomain involved in β_1–β_1 recognition. Given that the interaction between two β_1 subunits of the same species (dog–dog or rat–rat) is more effective than the interaction between rat and dog β_1 subunits [61], Tokhtaeva and colleagues [92] looked for surface-exposed species-specific amino acids in the sequence of β_1 subunit and identified four residues, which are different between both species and are contained in the 198–207 segment. Rat-like amino acid substitutions introduced in the dog β_1 subunit weakens its interaction with the endogenous dog β_1 subunit, whereas the insertion of the rat-specific Thr202 into the exogenous dog β_1 subunit impairs its interaction with the endogenous dog β_1 subunit to the level observed between dog and rat native subunits. The opposite effect is observed in the rat β_1 subunit upon the introduction of dog-like residues and the deletion of Thr202. These results suggest that the amino acid residues important for β_1–β_1 binding are located upstream and downstream of the Thr insertion position. The insertion or removal of the Thr residues in one of the two interacting subunits probably misaligns these binding residues, and thus causes the characteristic difference in affinity between the two species [92].

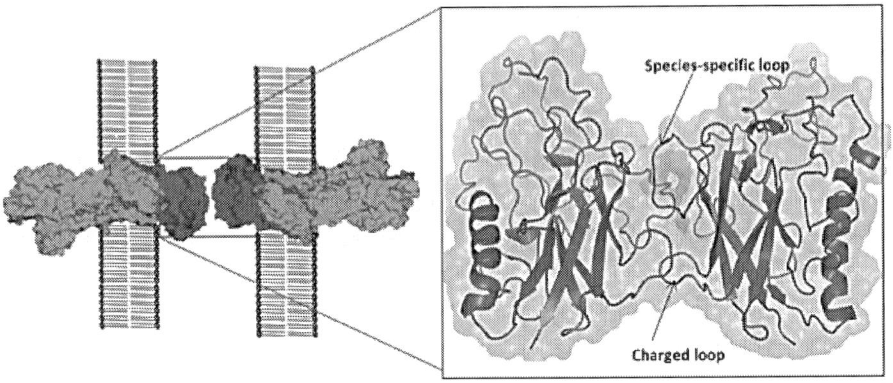

Figure 4. The intercellular adhesion between Na^+,K^+-ATPase β_1 subunit. A surface model based on the crystal structure illustrating the association of Na^+,K^+-ATPase dimer (α in green and β in blue) at the intercellular space. The magnified square shows one of the representative models resulting from the docking algorithm performed for the coupling of two Na^+,K^+-ATPase β_1 subunits structures obtained from the crystals. In this specific model, two loops form the core of the interaction, namely, the one containing the species-specific residues identified by [92] and the other comprised of an unusual sequence of eight consecutive charged residues (214KRDEDKDR221). This charged loop and other regions adjacent to the species-specific loop are suggested as the potential interface for β_1–β_1 interaction.

How specie-specific residues adjacent to Thr202 coordinate with residues residing at surrounding regions on the same β_1-subunit and with its interacting partner have yet to be elucidated. The segment 198–207 constitutes one of the characteristic protruding loops in the connecting β-strands B and C of the β_1-subunit extracellular fold. The majority of the ectodomain surface-exposed residues located most distal from the membrane reside within loops interconnecting β-strands, some of which must be involved in the dimer interface in conjunction with segment 198–207. Since the crystal structure of the Na^+,K^+-ATPase β_1 subunits now available [33;93], we modeled and predicted interacting surfaces on β-subunit and thus identified putative amino acids that participate in β_1–β_1 interaction. This approach will soon lead us to uncover a detailed adhesion mechanism, which is of great importance for epithelial physiology.

4. The Na^+,K^+-ATPase is the receptor of hormone ouabain

4.1. Cardiotonic steroids (CTSs)

The CSTs have been used for at least 200 years to treat heart failure and tachycardia due to their inotropic effect on the heart [94]. They are specific steroids and are extracted from plants of genus *Digitalis* and *Strophanthus* and from vertebrates such as several species of toads [95]. The CSTs have a steroid nucleus and can sort as cardenolides (with a five-membered lactone ring) or bufadienolides (six-membered lactone ring) and contain various combinations of hydroxyl, sulfate, or carbohydrates groups (Fig. 5) [96]. All types of CST bind with its receptor, the α-subunit of Na^+,K^+-ATPase, in a pocket formed by transmembrane segments M1–M6. The best affinity for the CST is of the E2P conformation [97]. The sensitivity of the sodium pump to CSTs is controlled by multiple elements mainly by the tissue specific distribution of α and β isoforms and by the glycosylation of CSTs. Thus, in the case of digoxin and digitoxin (Fig. 5), the affinity toward the Na^+,K^+-ATPase improves with up to fourfold preference for α_2/α_3 over α_1 isoforms [98].

Many studies have demonstrated the endogenous productions of CSTs in mammals. Thus, ouabain was detected in plasma [99], digoxin in urine [100], and marinobufagenin in plasma [101]. An interesting feature of Na^+,K^+-ATPase is the highly conserved nature of the CST-binding site, suggesting that this site plays a significant physiological role [94]. The normal ranges for circulating ouabain vary between 2500 ±500 pmol/l and 176 000 ± 68 000 pmol/l, depending on the measuring condition and the test used [102]. Interestingly, the binding of cardenolides and bufadienolides to the α-subunit of the Na^+,K^+-ATPase results not just in the inhibition of Na^+,K^+-ATPase ion transport activity but also in the activation of signaling cascades [103]. Moreover, endogenous ouabain is synthesized and secreted by the hypothalamus [104, 105] and the adrenocortical gland [106–107]. A status of hormone was recommended for the endogenous CSTs as it was demonstrated that it increases during exercise [108], salty meals [109–111], and pathological conditions such as arterial hypertension and myocardial infarction [112]. To confirm the hormone-like function, Arnaud-Batista and colleagues [113] showed that ouabain and bufalin induce diuresis, natriuresis, and kaliuresis, mediated by signal transduction in the isolated intact rat kidney. Furthermore, at the systemic level, cardenolides and bufadienolides have been implicated in many physiological and pathophysiological mechanisms, including cell growth and cancer, body or organ weight gain, mood disorders, vascular tone homeostasis, blood pressure, hypertension, and natriuresis[114].

4.2. The physiological role of hormone ouabain in epithelia

Fifteen years ago, the evidence that ouabain is a hormone was convincing enough as to start wondering what may its physiological role be. Our search was oriented by the observation that (MDCK) epithelial cells exposed to high concentrations of ouabain (≥1 μM) do not show sign of damage, but retrieve from the plasma membrane molecules involved in cell–cell and cell-substrate attachment, and detach from each other and from the substrate. These observations suggested that there is a mechanism that relates the occupancy of the pump (P) by ouabain to adhesion mechanisms (A). Accordingly, this mechanism was called P → A. We

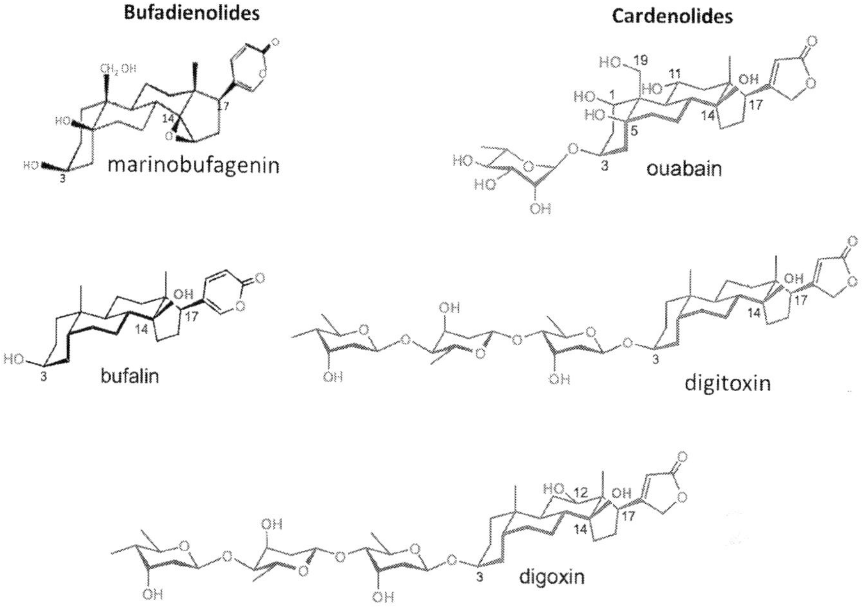

Figure 5. Structural features common for cardiotonic steroids (CTSs). All CTSs include a *cis–trans–cis* ring fused steroid core, which adopts a U-shaped conformation with a convex β-surface, a hydroxyl group at C14 (OH14β; purple). CTSs are classified as Cardenolides and Bufadienolides based on a five- or six-membered lactone ring in a β-conformation at position C17. Some CTSs have a carbohydrate moiety of one to four residues attached to C3. Ouabain, the most hydrophilic CTSs, is constituted of a steroid core with four hydroxyl groups at the β-surface (in blue), a hydroxyl group at the α-surface (purple), an unsaturated lactone ring of five members (green), and a rhamnose sugar moiety (pink). The structures of two members of the bufadienolides (marinobufagenin and bufalin) and three members of the cardenolides (ouabain, digitoxin, and digoxin) are illustrated.

discovered that P → A mechanism is associated with several signaling proteins such as cSrc and ERK1/2 (Fig. 6), and it consists of a loss of cytosolic K^+, an increase of cytosolic levels of Na^+ and Ca^{+2} and the activation of protein tyrosine kinases and ERK1/2. Ouabain binding also increases p190Rho-GAP, which enhances the GTPase activity of RhoA [115]. Detachment may not be ascribed to the ensuing decrease of K+ content because lowering the K-content by incubating the cells in media with only 0.1 mM K^+ (instead of the regular 4.0 mM) does not cause cell detachment [116].Therefore, we put forward the working hypothesis that ouabain at nanomolar concentrations, i.e., within the hormonal range in mammalian plasma, may act on the same junctional structures without provoking irreversible damages. To explore the plausibility of this idea, we experimentally tested the effect of ouabain on different cell–cell adhesion complexes starting with TJs. While toxic concentration of ouabain open the TJ, physiological concentrations of ouabain increase its hermeticity. Interestingly, the first effect depends on the pumping activity of Na^+,K^+-ATPase, whose inhibition perturbs the ionic balance of the cell. On the contrary, physiological concentrations of ouabain (i.e., in the nanomolar range) neither inhibit K^+ pumping nor disturb the K^+ balance of the cell [117]. At these concentrations, the effects of ouabain depends mainly on the activation of the receptor complex of Na^+,K^+-ATPase. While toxic levels of ouabain regulate the opening of TJs through endocytic and degradation processes, physiological concentrations of ouabain modulate TJs through changes in the molecular composition of the TJ through processes that provoke changes in transcription rate and expression of its proteins [118] (Fig. 6A). Another prominent cell–cell contact is the adherens junction (AJ) and one of the scaffolding proteins of this junction is β-catenin, a key member of the Wnt signaling pathway (Fig. 6B). During the activation of this pathway, β-catenin is translocated to the nucleus, where it modifies gene expression [119]. Interestingly, 10 nM and 1 μM ouabain provoke the translocation of β-catenin to the nucleus of MDCK cells [116]. Liu and co-workers [120] have recently found evidence that Na^+,K^+-ATPase, and E-cadherin are closely associated, indicating that E-Cadherin could be part of the signalosome of the Na^+,K^+-ATPase. To further explore the hypothesis that nanomolar concentrations of ouabain modulate cell–cell contacts, the effect of 10 nM ouabain have been studied on another type of cell–cell contact, the gap junction. In MDCK cells treated with this concentration of ouabain cell–cell communication have been increases by up to 510% in one hour. Moreover, inhibitors of transcription and of translation do not affect the induction of Gap junction communication (GJC) by ouabain, indicating that cells express a sufficient level of connexins to account for the rapid enhancement of GJC [121].

Ouabain effects through signaling were observed also in cardiac myocytes when nontoxic concentrations of ouabain, that partially inhibit the Na^+,K^+-ATPase, activate signaling pathways that regulate growth [122, 123]. Ouabain can activate signal cascades that vary between cell types, depending on the dose and the α-subunit isoform expressed in the cell [124, 103]. The existence of two pools of Na^+,K^+-ATPase within the plasma membrane with two distinct functions have been proposed: the classical ion pump whose partial inhibition by ouabain provokes an increase in $[Ca^{2+}]i$, and the second, the signal transducing pool which through protein–protein interactions regulates cell growth, proliferation, differentiation, and apoptosis. Part of the nontransporting Na^+,K^+-ATPase is located in the caveolae. cSrc is usually bound to the Na^+,K^+-ATPase in caveolae. Ouabain binding to the pump located in caveolae, stimulates cSrc activation, which consequently activates other downstream signaling pathways [125]. Signaling through Src is supported by the discovery that in a cell-free system, the addition of ouabain modifies the Na^+,K^+-ATPase cSrc complex and activates cSrc [128]. Alongside, the epidermal growth factor receptor (EGFR) is transactivated upon ouabain binding to Na^+,K^+-ATPase and additional signaling occurs that activate downstream targets including She, Grb, Ras, Raf, MEK, and ERK [125, 127] (Fig. 6). These signaling pathways regulate early response genes associated with cell growth and also regulate cell motility and a number of metabolic pathways [123,126]. Another signaling role was found by Aizman and coworkers [126]. In epithelial cells, the Na^+,K^+-ATPase interacts with the inositol 1,4,5-triphosphate receptor (IP3R) within the signaling microdomain. They show that interaction of ouabain with the signaling Na^+,K^+-ATPase provokes synchronized Ca^{2+} oscillations rising from the modification of such interaction. Those slow oscillations activate NF-kB.

http://dx.doi.org/10.5772/61244

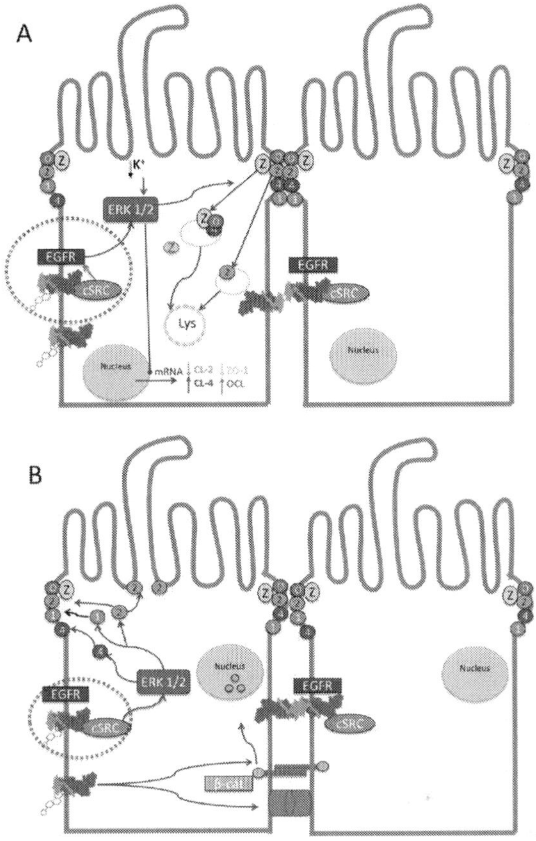

Figure 6. Signaling in the ouabain-induced modulation of cell contacts. Ouabain (red silhouette) induces the formation of a signalosome, a caveolar complex (discontinued grey line) including the Na+,K+-ATPase, its associated cSrc (cSRC) and the EGF receptor (EGFR). (A) Ouabain (300 nM) activates cSrc, which in turn transactivates the EGFR pathway, causing a phosphorylation of ERK1/2. The inhibition of the pump alters the ionic gradient that also contributes to the activation of ERK1/2. The activation of ERK1/2 is crucial for the clathrin- and dynamin-dependent endocytosis of TJ components. Two possible types of endocytic vesicles are formed: one containing a core complex with essential TJ proteins, such as ZO-1 (encircled Z); occludin (encircled O) and Claudin-4 (encircled 4) and a second one entailing components such as Claudin-2 (encircled 2) that makes TJs permeable to water and Na+. ERK1/2 activation is required to reduce the levels of Occludin, Claudin-4 and ZO-1 proteins, but not that of Claudin-2. ERK1/2 is also necessary to reduce Claudin-2 and ZO-1 mRNA levels. Notably, the cellular content of Claudin-4 and occludin mRNAs increases, during the opening of the TJs induced by Ouabain. (B) Epithelial cells treated with 10 nM of Ouabain (hormonal concentration) show increased tight junction sealing [117]. Activation of ERK 1/2 modulates the expression of Claudins (1,2 and 4) at the tight junction and promotes the expression of Claudin-2 in the cilium. Moreover, under this condition, cell-communication by gap junctions (red cylinder) is also increased by a mechanism still not well understood and β-catenin (khaki circles), a component of the Adherens junctions travels to the nucleus and modulates the expression of genes involved in cell-junction regulation.

Signal cascades vary between different cells types. For example, in cardiac myocytes and renal cell lines derived from the porcine kidneys (LLC-PK1) and the opossum kidneys (OK), ouabain-mediated activation of a signaling cascade has been demonstrated to be dependent upon the activation of Src, MAPK, and PI-3K pathways [129,122], whereas in human breast (BT20), prostate (DU145) cancer cells, and PY-17 cells, ouabain activate Src and MAPK pathway, but not PI-3K pathway [130]. Downstream in the ouabain-activated signaling cascade, the level of complexity increases due to the activation of several cell-specific secondary messengers and the cross talk between distinct pathways [131]. For each cell type, different pathways and branches are activated and only part of their complexity is known. In the MAPK pathway, several secondary messengers, downstream targets of ERK1/2, have been identified. Upon activation, ERK1/2 is able to migrate to the nucleus and activate several transcription factors (STAT1/3, c-fos, CREB, Elk-1) or in the cytoplasm modulate ion channels, receptors, or cytoskeleton proteins by direct phosphorylation.

5. Concluding remarks

Although without its β-subunit the Na^+,K^+-ATPase could not be expressed in the plasma membrane, nor have an enzymatic activity, no convincing role was detected for this subunit beyond of helping the α one to cage K^+. We have shown that, due to its adhesiveness, the β_1-subunit may establish a linkage with an identical subunit located in a neighboring cell across the intercellular space, and be thereby responsible for the polarized expression of Na^+,K^+-ATPase in epithelial cells. Furthermore, it has been demonstrated that β–β interaction stabilizes and maintains cell-junctions integrity in transporting epithelia. The molecular mechanism by which this interaction occurs is still far from being elucidated. Nevertheless, it is clear that both N-glycans and specific sequences exposed on the polypeptide surface are implicated. The observation that P → A mechanism is involved in the shuttling of β-catenin to the nucleus and thus in the Wnt/Wingless cascade, in the growth factor signaling pathways, as well as the ability of ouabain to enhance intercellular communication through gap junctions speaks of the important physiological role played by the hormone ouabain. The importance of this mechanism is compounded by the fact that in the meanwhile ouabain was shown to be a hormone that varies in response of several physiological and pathological conditions. Therefore, we may postulate that ouabain may determine the retrieval of the β-subunit from the plasma membrane and down regulates the expression of Na^+,K^+-ATPase in the cell membrane and thus, indirectly regulates the absorption and secretion of ions and nutrients. Therefore, ouabain should be added to the list of hormones that affect transepithelial transport, along with aldosterone, antidiuretic hormone, and the like.

Acknowledgements

Our experimental work was supported by the National Research Council of México (CONACYT). J. Lobato, T. López, O. Páez, and C. Vilchis were recipients of a Doctoral Fellowship from CONACYT-MEXICO.

http://dx.doi.org/10.5772/61244

Author details

Jorge Alberto Lobato Álvarez, Teresa del Carmen López Murillo,
Claudia Andrea Vilchis Nestor, María Luisa Roldán Gutierrez, Omar Páez Gómez and
Liora Shoshani*

*Address all correspondence to: shoshani@fisio.cinvestav.mx

Centro de Investigación y de EstudiosAvanzados del InstitutoPolitécnicoNacional (CINVES-TAV-IPN), Mexico city, Mexico

References

[1] Cereijido M, Contreras RG, Shoshani L. Cell adhesion, polarity, and epithelia in the dawn of metazoans. Physiol Rev. 2004; 84:1229–62.

[2] Cereijido M, Contreras RG, Shoshani L, Flores-Benitez D, Larre I. Tight junction and polarity interaction in the transporting epithelial phenotype. Biochim Biophys Acta. 2008;1778:770–93.

[3] Koefoed-Johnsen, V., and Ussing, H. H. The nature of the frog skin potential. Acta Physiol Scand 1958; 42, 298–308.

[4] Skou JC. The influence of some cations on an adenosine triphosphatase from peripheral nerves. Biochim Biophys Acta. 1957; 23:394–401.

[5] Schatzmann HJ. Cardiac glycosides as inhibitors of active potassium and sodium transport by erythrocyte membrane. Helv Physiol Pharmacol Acta 11. 1953; 346–354.

[6] Skou JC. Enzymatic basis for active transport of Na^+ and K^+ across cell membrane. Physiol Rev. 1965; 45:596–617.

[7] Garty H, Karlish SJ. Role of FXYD proteins ion transport. Annu Rev Physiol. 2006; 68:431–59.

[8] Morth JP, Pedersen BP, Toustrup-Jensen MS, Sørensen TL, Petersen J, Andersen JP, Vilsen B, Nissen P. Crystal structure of the sodium–potassium pump. Nature 2007; 450 1043–1049.

[9] Shinoda T, Ogawa H, Cornelius F, Toyoshima C. Crystal structure of the sodium-potassium pump at 2.4 Å resolution. Nature. 2009; 459 [7245]:446–50.

[10] Kanai R, Ogawa H, Vilsen B, Cornelius F, Toyoshima C. Crystal structure of a Na+-bound Na+,K+-ATPase preceding the E1P state. Nature. 2013; 502[7470]:201–6.

[11] Shull GE, Schwartz A, Lingrel JB. Amino-acid sequence of the catalytic subunit of the [Na+ + K+] ATPase deduced from a complementary DNA. Nature. 1985; 316:691–5.

[12] Kaplan JH. Biochemistry of Na,K-ATPase. Annu Rev Biochem. 2001; 71:511–35.

[13] Blanco G, Mercer RW. Isozymes of the Na-K-ATPase: heterogeneity in structure, diversity in function. Am. J. Physiol. 1998; 275: F633–F650.

[14] Jorgensen PL, Hakansson KO, Karlish SJ. Structure and mechanism of Na,K-ATPase: functional sites and their interactions. Annu Rev Physiol. 2003; 65:817–49.

[15] Colonna TE, Huynh L, Fambrough DM. Subunit interactions in the Na,K-ATPase explored with the yeast two-hybrid system. J Biol Chem. 1997; 272,12366–12372.

[16] Brotherus JR, Jacobsen L, Jørgensen PL. Soluble and enzymatically stable [Na++ K+]-ATPase from mammalian kidney consisting predominantly of Protomer αβ-units. Preparation, assay and reconstitution of active Na+,K+ transport. Biochim Biophys Acta. 1983; 731:290–303.

[17] Tamkun MM, Fambrough DM. The [Na+ + K+]-ATPase of chick sensory neurons. Studies on biosynthesis and intracellular transport. J Biol Chem. 1986; 261:1009–19.

[18] Miller RP, Farley RA. All three potential N-glycosylation sites of the dog kidney [Na+ + K+]-ATPase beta-subunit contain oligosaccharide. Biochim Biophys Acta. 1988; 954:50–7.

[19] Treuheit MJ, Costello CE, Kirley TL. Structures of the complex glycans found on the beta-subunit of [Na,K]-ATPase. J Biol Chem. 1993; 268:13914–9.

[20] Schmitz B, Peter-Katalinic J, Egge H, Schachner M. Monoclonal antibodies raised against membrane glycoproteins from mouse brain recognize N-linked oligomannosidic glycans. Glycobiolog. 1993; 3: 609–617.

[21] Chothia C, Jones EY. The molecular structure of cell adhesion molecules. Annu Rev Biochem. 1997; 66:823–62.

[22] Gloor S, Antonicek H, Sweadner JK, Pagliusi S, Frank R, Moos M, Schachner M. The adhesion molecule on glia [AMOG] is a homologue of the beta subunit of the Na,K-ATPase. J Cell Biol. 1990; 110: 165–174.

[23] Lutsenko S, Kaplan JH. An essential role for the extracellular domain of the Na^+,K^+-ATPase β-subunit in cation occlusion. Biochemistry. 1993; 32, 6737–6743.

[24] Geering K. The functional role of beta subunits in oligomeric P-type ATPases. J Bioenerg Biomembr. 2001; 33:425–38.

[25] Geering K. Functional roles of Na,K-ATPase subunits. Curr Opin Nephrol Hypertens. 2008;17:526-32

[26] Jones DH, Li TY, Arystarkhova E, Barr KJ, Wetzel RK, Peng J, Markham K, Sweadner KJ, Fong GH, Kidder GM. Na,K-ATPase from mice lacking the gamma subunit [FXYD2] exhibits altered Na+ affinity and decreased thermal stability. J Biol Chem. 2005; 280:19003–11.

[27] Goldschmidt I, Grahammer F, Warth R, Schulz-Baldes A, Garty H, Greger R, Bleich M. Kidney and colon electrolyte transport in CHIF knockout mice. Cell Physiol Biochem. 2004; 14:113–20.

[28] Apell HJ, Schneeberger A, Sokolov VS. Partial reactions of the Na,K-ATPase: kinetic analysis and transport properties. Acta Physiol Scand Suppl. 1998; 643:235–45.

[29] Holmgren M, Wagg J, Bezanilla F, Rakowski RF, De Weer P, Gadsby DC. Three distinct and sequential steps in the release of sodium ions by the Na+/K+-ATPase. Nature. 2000; 403(6772):898–901.

[30] Rakowski RF, Gadsby DC, De Weer P. Voltage dependence of the Na/K pump.J Membr Biol. 1997; 155(2):105–12.

[31] Albers RW. Biochemical aspects of active transport. Annu Rev Biochem. 1967; 36:727–56.

[32] Albers RW, Koval GJ. Sodium-potassium-activated adenosine triphosphatase. VII. Concurrent inhibition of NA + -K + -adenosine triphosphatase and activation K + -nitrophenylphosphatase activities. J Biol Chem. 1972; 247(10):3088–92.

[33] Nyblom M, Poulsen H, Gourdon P, Reinhard L, Andersson M, Lindahl E, Fedosova N, Nissen P. Crystal structure of Na+,K[+]-ATPase in the Na[+]-bound state. Science 2013; 342:123–7.

[34] Kanai R, Ogawa H, Vilsen B, Cornelius F, Toyoshima C. Crystal structure of a Na+-bound Na+,K+-ATPase preceding the E1P state. Nature. 2013; 502(7470):201–6.

[35] Cereijido M, Contreras RG, Shoshani L, García-Villegas MR. Membrane targeting. Prog Biophys Mol Biol. 2003; 81:81–115.

[36] Gonzalez A, Rodriguez-Boulan E. Clathrin and AP1B: key roles in basolateral trafficking through trans-endosomal routes. FEBS Lett. 2009; 583(23):3784–95.

[37] Matter K, Yamamoto EM, Mellman I. Structural requirements and sequence motifs for polarized sorting and endocytosis of LDL and Fc receptors in MDCK cells. J Cell Biol. 1994; 126(4):991–1004.

[38] Thomas DC, Brewer CB, Roth MG. Vesicular stomatitis virus glycoprotein contains a dominant cytoplasmic basolateral sorting signal critically dependent upon a tyrosine. J Biol Chem. 1993; 268(5):3313–20.

[39] Caplan MJ, Anderson HC, Palade GE, Jamieson JD. Intracellular sorting and polarized cell surface delivery of [Na+,K+]ATPase, an endogenous component of MDCK cell basolateral plasma membranes. Cell. 1986; 46[4]:623–3.

[40] Mays RW, Siemers KA, Fritz BA, Lowe AW, van Meer G, Nelson WJ. Hierarchy of mechanisms involved in generating Na/K-ATPase polarity in MDCK epithelial cells. J Cell Biol. 1995; 130[5]:1105–15.

[41] Blostein R, Zhang R, Gottardi CJ, Caplan MJ. Functional properties of an H,K-ATPase/Na,K-ATPase chimera. J Biol Chem. 1993; 268:10654–8.

[42] Gottardi CJ, Caplan MJ. Molecular requirements for the cell-surface expression of multisubunit ion-transporting ATPases. Identification of protein domains that participate in Na,K-ATPase and H,K-ATPase subunit assembly. J Biol Chem. 1993b; 268:14342–7.

[43] Muth TR, Gottardi CJ, Roush DL, Caplan MJ. A basolateral sorting signal is encoded in the alpha-subunit of Na-K-ATPase. Am. J. Physiol.1998; 274: C688–C696.

[44] Dunbar LA, Aronson P, Caplan MJ. A transmembrane segment determines the steady-state localization of an ion-transporting adenosine triphosphatase. J Cell Biol. 2000; 14:769–78.

[45] Bonifacino JS. Adaptor proteins involved in polarized sorting. J Cell Biol. 2014; 204:7–17.

[46] Duffield A, Fölsch H, Mellman I, Caplan MJ. Sorting of H,K-ATPase beta-subunit in MDCK and LLC-PK cells is independent of mu 1B adaptin expression. Traffic. 2004;5[6]:449–61.

[47] Gravotta D, Deora A, Perret E, Oyanadel C, Soza A, Schreiner R, Gonzalez A, Rodriguez-Boulan E. AP1B sorts basolateral proteins in recycling and biosynthetic routes of MDCK cells. Proc Natl Acad Sci U S A. 2007; 104: 1564–1569.

[48] Farr GA, Hull M, Mellman I, Caplan MJ. Membrane proteins follow multiple pathways to the basolateral cell surface in polarized epithelial cells. J Cell Biol. 2009; 186:269–82.

[49] Nelson WJ, Hammerton RW, McNeill H. Role of the membrane-cytoskeleton in the spatial organization of the Na,K-ATPase in polarized epithelial cells. Soc Gen Physiol Ser. 1991; 46: 77–87.

[50] Nusrat A, Turner JR, Madara JL. Molecular physiology and pathophysiology of tight junctions. IV. Regulation of tight junctions by extracellular stimuli: nutrients, cytokines, and immune cells. Am J Physiol Gastrointest Liver Physiol. 2000; 279(5):G851–7.

[51] Nusrat A, Brown GT, Tom J, Drake A, Bui TT, Quan C, Mrsny RJ. Multiple protein interactions involving proposed extracellular loop domains of the tight junction protein occluding. MolBiol Cell. 2005; 16(4):1725–34.

[52] Geering K, Theulaz I, Verrey F, Häuptle MT, Rossier BC. A role for the beta-subunit in the expression of functional Na+-K+-ATPase in Xenopus oocytes. Am J Physiol. 1989; 257:C851–8.

[53] Ackermann U, Geering K. Mutual dependence of Na,K-ATPase alpha- and beta-subunits for correct posttranslational processing and intracellular transport. FEBS Lett. 1990; 269:105–8.

[54] Noguchi S, Mishina M, Kawamura M, Numa S. Expression of functional [Na+ + K+]-ATPase from cloned cDNAs. FEBS Lett. 1987; 225:27–32.

[55] Fambrough DM. The sodium pump becomes a family. Trends Neurosci. 1988;11(7): 325–8.

[56] Takeyasu K, Kawakami K. Na+,K(+)-ATPase: genes, expression and membrane insertion]. Seikagaku. 1989; 61(5):394–401.

[57] Horowitz B, Eakle KA, Scheiner-Bobis G, Randolph GR, Chen CY, Hitzeman RA, Farley RA. Synthesis and assembly of functional mammalian Na,K-ATPase in yeast. J Biol Chem. 1990; 265(8):4189–92.

[58] Cereijido M, Ehrenfeld J, Meza I, Martínez-Palomo A. Structural and functional membrane polarity in cultured monolayers of MDCK cells. J Membr Biol. 1980; 52(2): 147–59.

[59] Contreras RG, Lázaro A, Bolivar JJ, Flores-Maldonado C, Sánchez SH, González-Mariscal L, García-Villegas MR, Valdés J, Cereijido M. A novel type of cell-cell cooperation between epithelial cells. J Membr Biol. 1995; 145(3):305–10.

[60] Shoshani L, Contreras RG, Roldan ML, Moreno J, Lazaro A, Balda MS, Matter K, Cereijido M. The polarized expression of Na+,K+-ATPase in epithelia depends on the association between beta-subunits located in neighboring cells. Mol Biol Cell. 2005; 16: 1071–1081.

[61] Padilla-Benavides T, Roldan ML, Larre I, Flores-Benitez D, Villegas-Sepulveda N, Contreras RG, Cereijido M, Shoshani L. The polarized distribution of Na^+,K^+-ATPase: role of the interaction between beta subunits. Mol. Biol. Cell. 2010; 21: 2217–2225.

[62] Vagin O, Tokhtaeva E, Sachs G. The role of the beta 1 subunit of the Na,K-ATPase and its glycosylation in cell-cell adhesion. J Biol Chem. 2006; 281: 39573–39587.

[63] Rajasekaran SA, Barwe SP, Gopal J, Ryazantsev S, Schneeberger EE, Rajasekaran AK. Na-K-ATPase regulates tight junction permeability through occludin phosphorylation in pancreatic epithelial cells. Am J Physiol Gastrointest Liver Physiol. 2007; 292: G124–133.

[64] Nelson WJ, Veshnock PJ. Ankyrin binding to [Na^+ + K^+] ATPase and implications for the organization of membrane domains in polarized cells. Nature.1987; 328: 533–536.

[65] Vagin O, Tokhtaeva E, Yakubov I, Shevchenko E, Sachs G. Inverse correlation between the extent of N-glycan branching and intercellular adhesion in epithelia. Contribution of the Na,K-ATPase beta1 subunit. J Biol Chem. 2008; 283: 2192–2202.

[66] Rajasekaran SA, Palmer LG, Moon SY, Peralta-Soler A, Apodaca GL, Harper JF, Zheng Y, Rajasekaran AK. Na,K-ATPase activity is required for formation of tight

junctions, desmosomes, and induction of polarity in epithelial cells. Mol Biol Cell. 2001; 12: 3717–3732.

[67] Tokhtaeva E, Sachs G, Souda P, Bassilian S, Whitelegge JP, Shoshani L, Vagin O. Epithelial junctions depend on intercellular trans-interactions between the Na,K-ATPase β_1 subunits. J Biol Chem. 2011; 286: 25801–25812.

[68] Genova JL, Fehon RG. Neuroglian, Gliotactin, and the Na+/K+ ATPase are essential for septate junction function in Drosophila. J Cell Biol. 2003;161:979–989.

[69] Paul SM, Palladino MJ, Beitel GJ. A pump-independent function of the Na,K-ATPase is required for epithelial junction function and tracheal tube-size control. Development [Cambridge, England]. 2007; 134, 147–155.

[70] Antonicek H, Persohn E, Schachner M. Biochemical and functional characterization of a novel neuron-glia adhesion molecule that is involved in neuronal migration. J Cell Biol. 1987 104: 1587–1595.

[71] Antonicek H, Schachner M. The Adhesion Molecule on Glia [AMOG] Incorporated into lipid vesicles binds to subpopulations of neurons. J Neurosci. 1988; 8 [8]: 2961–66.

[72] Martin-Vasallo P, Dackowski W, Emanuel JR, Levenson R. Identification of a putative isoform of the Na,K-ATPase beta subunit. Primary structure and tissue-specific expression. J Biol Chem 2. 1989; 64: 4613–4618.

[73] Müller-Husmann G, Gloor S, Schachner M. Functional characterization of β isoforms of murine Na,K-ATPase the adhesion molecule on glia [AMOG/β_2], but not β_1, promotes neurite outgrowth. J Biol Chem. 1993; 268 [35]: 26260–26267.

[74] Kirk P, Wilson MC, Heddle C, Brown MH, Barclay AN, Halestrap AP. CD147 is tightly associated with lactate transporters MCT1 and MCT4 and facilitates their cell surface expression. EMBO J. 2000; 19: 3896–3904.

[75] Philp NP, Ochrietor JD, Rudoy C, Muramatsu T, Linser PJ. Loss of MCT1, MCT3 and MCT4 expression in the retinal pigment epithelium and neural retina of the 5A11/Basigin null mouse. Invest. Opthalmol. Vis. Sci. 2003; 44: 1305–1311.

[76] Wilson MC, Meredith D, Fox JE, Manoharan C, Davies AJ, Halestrap AP. Basigin [CD147] is the target for organomercurial inhibition of monocarboxylate transporter isoforms 1 and 4: the ancillary protein for the insensitive MCT2 is EMBIGIN [gp70]. J Biol Chem. 2005; 280: 27213–27221.

[77] Deora AA, Philp N, Hu J, Bok D, Rodriguez-Boulan E. Mechanisms regulating tissue-specific polarity of monocarboxylate transporters and their chaperone CD147 in kidney and retinal epithelia. PNAS. 2005; 102 [45]: 16245–16250.

[78] Heller M, von der Ohe M, Kleene R, Mohajeri MH, Schachner M. The immunoglobulin-superfamily molecule basigin is a binding protein for oligomannosidic carbohydrates: an anti-idiotypic approach. J Neurochem. 2003; 84: 557–565.

[79] Kleene R, Loers G, Langer J, Frobert Y, Buck F, Schachner M. Prion protein regulates glutamate-dependent lactate transport of astrocytes. J Neurosci. 2007 27[45]: 12331–12340.

[80] Dunbar LA, Roush DL, Courtois-Coutry N, Muth TR, Gottardi CJ, Rajendran V, Geibel J, Kashgarian M, Caplan MJ. Sorting of ion pumps in polarized epithelial cells. Ann N Y Acad Sci. 1997; 834:514–23.

[81] Wilson PD, Devuyst O, Li X, Gatti L, Falkenstein D, Robinson S, Fambrough D, Burrow CR. Apical plasma membrane mispolarization of Na,K-ATPase in polycystic kidney disease epithelia is associated with aberrant expression of the beta2 isoform. Am. J. Pathol. 2000; 156: 253–68.

[82] Wilson PD. Polycystic Kidney Disease. N Engl J Med. 2004 350: 151–64.

[83] Mobasheri A, Oukrif D, Dawodu SP, Sinha M, Greenwell P, Stewart D, Djamgoz MB, Foster CS, Martin-Vasallo P, Mobasheri R. Isoforms of Na+,K+-ATPase in human prostate; specificity of expression and apical membrane polarization. Histol Histopathol. 2001; 16: 141–54.

[84] Ruiz A, Bhat SP, Bok D. Expression and synthesis of the Na,K-ATPase beta 2 subunit in human retinal pigment epithelium Gene. 1996; 176: 237–42.

[85] Vagin O, Turdikulova S, Sachs G. The H, K-ATPase beta subunit as a model to study the role of N-glycosylation in membrane trafficking and apical sorting. J Biol Chem. 2004 279: 39026–34.

[86] Mochizuki K, Kagawa T, Numari A, Harris MJ, Itoh J, Watanabe N, Mine T, Arias IM. Two N-linked glycans are required to maintain the transport activity of the bile salt export pump [ABCB11] in MDCK II cells. Am J Physiol Gastrointest Liver Physiol. 2007; 292: G818–28.

[87] Martinez-Maza R, Poyatos I, Lopez-Corcuera B, Nu E, Gimenez C, Zafra F, Aragon C. The role of N-glycosylation in transport to the plasma membrane and sorting of the neuronal glycine transporter GLYT2. J Biol Chem. 2001; 276: 2168–73.

[88] Zafra F, Gimenez C. Molecular determinants involved in the asymmetrical distribution of glycine transporters in polarized cells. Biochem Soc Trans. 2001; 29: 746–750.

[89] Gut A, Kappeler F, Hyka N, Balda MS, Hauri HP, Matter K. Carbohydrate-mediated Golgi to cell surface transport and apical targeting of membrane proteins. EMBO J. 1998; 17: 1919–1929.

[90] Vagin O, Turdikulova S, Sachs G. Recombinant addition of N-glycosylation sites to the basolateral Na,K-ATPase beta1 subunit results in its clustering in caveolae and apical sorting in HGT-1 cells. J Biol Chem. 2005; 280: 43159–43167.

[91] Bab-Dinitz E, Albeck S, Peleg Y, Brumfeld V, Gottschalk KE, Karlish SJ. A C-terminal lobe of the beta subunit of Na+,K+-ATPase and H+,K+-ATPase resembles cell adhesion molecules. Biochemistry. 2009; 15;48[36]:8684–91.

[92] Tokhtaeva E, Sachs G, Sun H, Dada LA, Sznajder JI, Vagin O. Identification of the amino acid region involved in the intercellular interaction between the β_1 subunits of Na+/K+ -ATPase. J Cell Sci. 2012; 125[Pt 6]:1605–16.

[93] Toyoshima C, Kanai R, Cornelius F. First crystal structures of Na+,K+-ATPase: new light on the oldest ion pump. Structure.2011; 19(12):1732–8.

[94] Shoner W, Bauer N, Müller-Ehmsen J, Krämer U, Hambarchian N, Schwinger R, Moeller H, Kost H, Weitkamp C, Scheitzer T, Kirch U, Neu H, Ernest-Güther Grünbaum. Ouabain as a Mammalian hormone. Ann N Y Acad Sci. 2003; 988:678–84.

[95] Lichtstein D, Rosen H, Dvela M. Cardenolides and bufadienolides as hormones: what is missing? Am J Physiol Renal Physiol. 2012; 302(8): F957–58.

[96] Cornelius F, Mahmmoud YA. Functional modulation of the sodium pump: the regulatory proteins "Fixit" News Physiol. Science 2009; 18 (3) 119–24.

[97] Laursen M, Gregersen JL, Yatime L, Nissen P, Fedosova NU. Structures and characterization of digoxin- and bufalin-bound Na+,K+-ATPase compared with the ouabain-bound complex. PNAS. 2015; 112 (6) 1755–60.

[98] Katz A. Selectivity of digitalis glycosides for isoforms of human Na,K-ATPase. J Biol Chem. 2010; 285(25):19582–92.

[99] Hamlyn JM, Blaustein MP, Bova S, DuCharme DW, Harris DW, Mandel F. Identification and characterization of a ouabain-like compound from human plasma. Proc Natl Acad Sci. 1991; 88, 6259–63.

[100] Goto A, Ishiguro T, Yamada K, Ishii M, Yoshioka M, Eguchi C. Isolation of a urinary digitalis-like factor indistinguishable from digoxin. Biochem Biophys Res Commun. 1990; 173, 1093–1.

[101] Bagrov AY, Fedorova OV, Austin-Lane JL, Dmitrieva RI, Anderson DE. Endogenous marinobufagenin-like immunoreactive factor and Na+,K+-ATPase inhibition during voluntary hypoventilation. Hypertension 1995a; 26, 781–8.

[102] Nicholls MG, Lewisa LK, Yandlea TG, Lordb G, McKinnonc W, Hiltonc P. Ouabain, a circulating hormone secreted by the adrenals, is pivotal in cardiovascular disease. Fact or fantasy?. Journal of Hypertension 2009; 27:3–8.

[103] Xie Z, Askari A. Na+,K+-ATPase as a signal transducer. Eur. J. Biochem. 2002; 269, 2434–39.

[104] Ludens JH, Clark MA, Robinson FG, DuCharme DW. Rat adrenal cortex is asource of a circulating ouabainlike compound. Hypertension. 1992; 19, 721–4.

[105] El-Masri MA, Clark BJ, Qazzaz HM, Valdes RJr. Human adrenal cells in culture produce both ouabain-like and dihydro ouabain-like factors. Clin Chem. 2002; 48, 1720–30.

[106] Doris PA, Hayward-Lester A, Bourne D, Stocco DM. Ouabain production by cultured adrenal cells. Endocrinology.1996; 137:2, 533–9.

[107] Qazzaz HM, Cao Z, Bolanowski DD, Clark BJ, Valdes RJr. De novo biosynthesis and radiolabeling of mammalian digitalis-like factors. Clinical Chemestry. 2004; 50(3): 612–20.

[108] Bauer N, Müller-Ehmsen J, Krämer U, Hambarchian N, Zobel C, Schwinger RH, Neu H, Kirch U, Grünbaum EG, Schoner W. Ouabain-like compound changes rapidly on physical exercise in humans and dogs: effects of beta-blockade and angiotensin-converting enzyme inhibition. Hypertension. 2005; 45[5]:1024–8.

[109] Fedorova OV, Agalakova NI, Talan MI, Lakatta EG, Bagrov AY. Brainouabain stimulates peripheral marinobufagenin via angiotensin II signalling in NaCl-loadedDahl-S rats. J Hypertens. 2005a; 23, 1515–23.

[110] Fedorova OV, Kolodkin NI, Agalakova NI, Namikas AR, Bzhelyansky A, St-Louis J, Lakatta EG, Bagrov AY. Antibody to marinobufagenin lowers blood pressure inpregnant rats on a high NaCl intake. J Hypertens. 2005b; 23, 835–42.

[111] Fedorova OV, Lakatta EG, Bagrov AY. Endogenous Na,K pump ligands aredifferentially regulated during acute NaCl loading of Dahl rats. Circulation. 2000; 102, 3009–14.

[112] Manunta P, Ferrandi M, Bianchi G, Hamlyn J M.Endogenous ouabain in cardiovascular function and disease. J Hypertens. 2009; 27 (1); 9–18.

[113] Arnaud-Batista FJ, Costa GT, Oliveira IM, Costa PP, Santos CF, Fonteles MC, Uchôa DE, Silveira ER, Cardi BA, Carvalho KM, Amaral LS, Pôças ES, Quintas LE, Noël F, Nascimento NR. Natriuretic effect of bufalin in isolated rat kidneys involves activation of the Na+-K+-ATPase-Src kinase pathway. Am J Physiol Renal Physiol. 2012; 302(8):F959–66.

[114] Nesher M, Shpolansky U, Rosen H, Lichtstein D. The digitalis-like steroid hormones: new mechanisms of action and biological significance. Life Science. 2007; 80(23): 2093–107.

[115] Contreras RG, Shoshani L, Flores-Maldonado C, Lazaro A, Cereijido M. Relationship between Na+,K+-ATPase and cell attachment. J Cell Sci 1999; 112:4223–32.

[116] Contreras RG, Flores-Maldonado C, Lazaro A, Shoshani L, Flores-Benitez D, Larre I, Cereijido M. Ouabain binding to Na^+,K^+ -ATPase relaxes cell attachment and sends a specific signal [NACos] to the nucleus. J Membr Biol. 2004; 198: 147–58.

[117] Larre I, Lazaro A, Contreras RG, Balda MS, Matter K, Flores-Maldonado C, Ponce A, Flores-Benitez D, Rincon-Heredia R, Padilla-Benavides T, Castillo A, Shoshani L, Cereijido M. Ouabain modulates epithelial cell tight junction. Proc Natl Acad Sci U S A. 2010; 107(25):11387–92.

[118] Rincon-Heredia R, Flores-Benitez D, Flores-Maldonado C,Bonilla-Delgado J, Garciá-Hernández V, Verdejo-Torres O, Castillo A, Larré I, Poot-Hernández CA, Franco M, Gariglio P, Reyes J, Contreras R.G. Ouabain induces endocytosis and degradation of tight junction proteins through ERK1/2-dependent path ways. Experimental Cell Research. 2014; 320 (1): 108–18.

[119] Heuberger J, Birchmeier W. Interplay of Cadherin-Mediated Cell Adhesion and Canonical Wnt Signaling. Cold Spring Harbor Perspectives in Biology. 2010; 2[2], a002915.

[120] Liu L, Ivanov AV, Gable ME, Jolivel F, Morrill GA, Askari A. Comparative properties of caveolar and noncaveolar preparations of kidney Na+/K+-ATPase. Biochemistry. 2011; 50[40]:8664–73.

[121] Ponce A, Larre L, Castillo A, García-Villegas R, Romero A, Flores-Maldonado C, Martinez- Redón J, Contreras R.G. Cereijido M. Ouabain Increases Gap Junctional Communication in Epithelial Cells. Cell Physiol Biochem. 2014; 34:2081–90.

[122] Liu J, Liang M, Liu L, Malhotra D, Xie Z, Shapiro JI. Ouabain-induced endocytosis of the plasmalemmal Na+,K+-ATPase in LLC-PK1 cells requires Caveolin-1. Kidney Int. 2005; 67, 1844–54.

[123] Peng M, Huang L, Xie Z, Huang WH, Askari A. Partial inhibition of Na^+,K^+-ATPase by ouabain induces the Ca^{2+}-dependent expressions of early-response genes in cardiac myocytes. J Biol Chem. 1996; 271, 10372–8.

[124] Pierre SV, Sottejeau Y, Gourbeau JM, Sánchez G, Shidyak A, Blanco G. Isoform-specificity of Na,K-ATPase-mediated ouabain signaling. Am J Physiol Renal Physiol. 2008;294, F859–66.

[125] Pierre SV, Xie Z. The Na+,K+-ATPase receptor complex: its organization and membership. Cell Biochem Biophys. 2006, 46, 303–16.

[126] Aizman O, Uhlen P, Lal M, Brismar H, Aperia A. Ouabain, a steroid hormone that signals with slow calcium oscillations. Proc Natl Acad Sci. 2001; 98, 13420–24.

[127] Haas M, Wang H, Tian J, Xie Z. Src-mediated inter-receptor cross-talk between the Na,K-ATPase and the epidermal growth factor receptor relays the signal from ouabain to mitogen-activated protein kinases. J Biol Chem. 2002; 277, 18694–702.

[128] Tian J, Cai T, Yuan Z, Wang H, Li L, Haas M. Binding of Src to Na+,K+ -ATPase forms a functional signaling complex. Mol Biol Cell. 2006; 17, 317–26.

[129] Khundmiri SJ, Amin V, Henson J, Lewis J, Ameen M, Rane MJ, Delamere N. A. Ouabain stimulates protein kinase B (Akt) phosphorylation in opossum kidney proximal tubule cells. American Journal of Physiology—Cell Physiology. 2007; 293 (3) C1171–80.

[130] Tian J, Li X, Liang M, Liu L, Xie JX, Ye Q. Changes in sodium pump expression dictate the effects of ouabain on cell growth. J Biol Chem. 2009; 284, 14921–9.

[131] Wang H, Haas M, Liang M, Cai T, Tian J, Li S. Ouabain assembles signaling cascades through the caveolar Na,K-ATPase. J. Biol. Chem. 2004; 279, 17250–5.

Chapter 3

Cell-cycle Alterations in Post-mitotic Cells and Cell Death by Mitotic Catastrophe

Duccio Lombardi and Laura Lasagni

Additional information is available at the end of the chapter

http://dx.doi.org/10.5772/61783

Abstract

Mitotic catastrophe (MC) has long been accounted as a cell death path activated by premature or inappropriate entry of cells into mitosis following chemical or physical stresses. Although various possible explanations related to MC have been formulated, no general accepted definition of this phenomenon has been found yet. Recent evidences, however, demonstrate that MC is not a distinguished way of cell death, rather a "pre-stage" anticipating cell death, taking place in mitotically disrupted cells, which later occurs via necrosis or apoptosis. Moreover, even though it is widely accepted that MC is the main outcome after ionizing radiation treatment or treatment with drugs that influence microtubule assembly/stability inducing mitotic failure, the final cell death pathway and the final outcome of MC, which strongly depend on the cell type and its related molecular profile, still need to be fully elucidated. Post-mitotic cells, like neurons in the central nervous system, and podocytes or tubular cells in the kidney, are particularly susceptible to MC. In the central nervous system, MC has been claimed as the cause of neuronal death in many neurologic disorders, while MC in podocytes and tubular death is connected with the development of progressive glomerulosclerosis.

Keywords: Mitotic catastrophe, neuron, podocyte, tubular cell, cell-cycle reentry

1. Introduction

Cell cycle is as old as life itself. In most situations, it is a generative force that creates new cells from old. It is a tightly regulated process whose misregulation can lead to unchecked proliferation and neoplastic disease. Moreover, a decade ago, it was hypothesized that cell-cycle abnormalities may be intimately connected with the death of terminally differentiated cells, such as neurons. In this case, the consequence of cell-cycle alterations is loss of cells, and this phenomenon has been postulated as a mechanism of pathogenesis in several neurodegenerative disorders. Recently, the link between aberrant cell-cycle reentry, cell death, and degenerative diseases has been observed also in other post-mitotic cell types, such as podocytes and tubular cells in the kidney. In this kind of process, post-mitotic cells enter into the cell cycle in response to stress signals in order to substitute death cells, but the absence or malfunction of a specific array of cell-cycle proteins may not allow for its completion. The final result is that the cells can neither reverse the course of the cell cycle or complete division, remaining locked in a non-functional state that push them to trigger a programmed cell death response. Interestingly, in *in vitro* and animal models of neurodegenerative diseases, the presence of active apoptotic pathways has been observed and reported, while there are conflicting data on the activation of classic apoptotic pathways in the human damaged tissues. Thus, it remains possible that the cell-cycle-linked cell death response may occur through different still not completely understood pathways. Mitotic catastrophe (MC) has long been accounted as a cell death path activated by premature or inappropriate entry of cells into mitosis following chemical or physical stresses. Although various possible explanations related to MC have been formulated, no generally accepted definition of this phenomenon has been found yet. Recent evidence, however, demonstrate that MC is not a distinguished way of cell death, rather a pre-stage anticipating cell death, taking place in mitotically disrupted cells, which later occurs via necrosis or apoptosis. Moreover, even though it is widely accepted that MC is the main outcome after ionizing radiation treatment or treatment with drugs that influence microtubule assembly/stability inducing mitotic failure, the final cell death pathway and the final outcome of MC, which strongly depend on the cell type and its related molecular profile, still need to be fully elucidated. Post-mitotic cells, like neurons in the central nervous system, and podocytes in the kidney, are particularly susceptible to MC. In the central nervous system, MC has been claimed as the cause of neuronal death in many neurologic disorders, while MC in podocytes is connected with the development of progressive glomerulosclerosis.

2. Regulation of the cell cycle

The cell cycle of eukaryotic cells comprises four main successive phases: G1 phase (first gap), S phase (DNA synthesis), G2 phase (second gap), and M phase (mitosis) (Figure 1). The orderly transition from one phase to the following and subsequent progression through the mitotic cycle is controlled by a group of protein kinases whose activity is central to this process, the cyclin-dependent kinases (CDKs). Their levels in the cell remain fairly stable, but each must bind with their activating partners, cyclins, whose levels of expression fluctuate throughout the cycle.

Mitogenic signals, such as soluble growth factors or cell-to-cell contact, stimulate the activation of D-type cyclins and their connection with CDK4 or CDK6. Cyclin D-CDK4 and cyclin D-CDK-6 complexes phosphorylate the retinoblastoma protein (Rb) and inhibit its affinity to bind the transcription factor E2F-1. Thus, E2F-1 is free to induce the transcription of specific genes involved in DNA replication. Moreover, in late G1, inhibition of Rb activates the expression of cyclin E that binds with CDK2. The cyclin E-CDK2 complex ensures the G1/S transition to occur by fully inactivating Rb by hyperphosphorylation. Thus, CDK2 to regulate progression

http://dx.doi.org/10.5772/61783

from G1 into S phase. Cyclin A binds with CDK2 that phosphorylates various substrates allowing DNA replication. The formation of cyclin A/CDK2 complex is required during S phase. After completion of S phase, DNA replication ceases and cells enter the G2 phase of the cycle. The cyclin A-CDK1 complex plays a central role in the transition from S to G2/M phase of the cell cycle by regulating the phosphorylation of specific substrates necessary for the completion of the G2 and M phases of the cell cycle. Mitosis is further regulated by cyclin B-CDK1 complex, which appears in late G2 and triggers the G2/M transition. Cyclin A is degraded and the system is reset. In this way, cells are now ready to start a new cell cycle when the presence of mitogenic stimuli induces the upregulation of D-type cyclins.

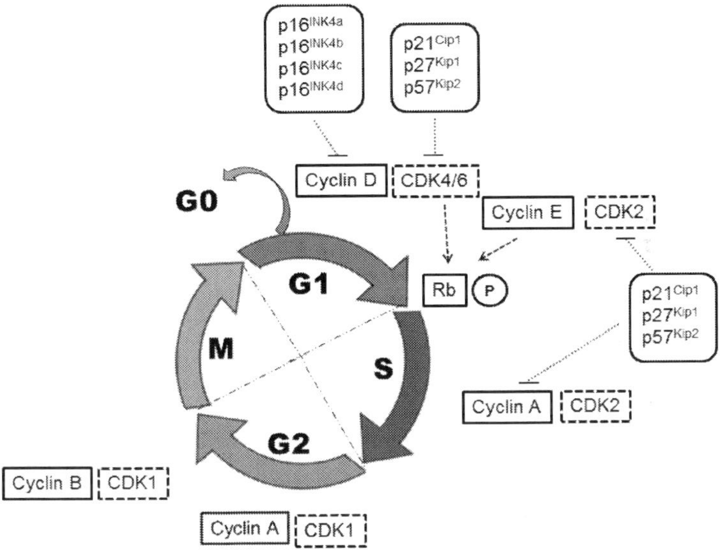

Figure 1. Schematic representation of the eukaryotic cell cycle.

CDK activity can be counteracted through post-translational modifications and subcellular translocations of specific CDK inhibitors (CDKIs), which bind with CDK alone or to the CDK–cyclin complex. CDKIs are organized in two families: INK4 and Cip/Kip. The INK4 family (inhibitors of cyclin D-dependent kinases) includes four members—$p16^{INK4a}$, $p15^{INK4b}$, $p18^{INK4c}$, and $p19^{INK4d}$—which specifically inactivate G1 CDK (CDK4 and CDK6) and the Cip/Kip family (inhibitors of cyclin D-, cyclin E-, and cyclin A-dependent kinases) comprises $p21^{Cip1}$, $p27^{Kip1}$, and $p57^{Kip2}$. CDKIs are regulated by both internal and external signals. The intracellular localization of different cell-cycle-regulating proteins also contributes to a correct cell-cycle progression.

Two important checkpoints (G1/S and G2/M) coordinate CDKs activity and ensure that each stage of the cell cycle is correctly completed before allowing further progress through the cycle.

If conditions are inadequate, the cell will not be allowed to progress through the cell cycle and be both arrested, until conditions are favorable, or induced to die through apoptosis (reviewed in reference [2]). The G1/S checkpoint, also known as the restriction point in mammalian cells, is defined as a point of no return in G1, following which the cell is committed to enter the cell cycle. It is necessary to control the progression of cell cycle in the presence of DNA damage. At this checkpoint, p53 activity arrests cell cycle induced by DNA damage, stimulating the transcription of different genes including p21. At the G2/M checkpoint, mitotic entry is prevented in response to DNA damage by mechanisms similar to those in the G1/S checkpoint. An additional checkpoint, the mitotic spindle checkpoint, occurs at the point in metaphase where all the chromosomes should have aligned at the mitotic plate and be under bipolar tension.

3. Molecular basis of mitotic catastrophe

"Mitotic catastrophe" has been reported, for the first time, in a temperature-sensitive lethal phenotype of *Schizosaccharomyces pombe* in 1989.[3] The first hallmark that induced researchers to distinguish this cell-death modality from others already characterized (i.e., apoptosis) was the macroscopic alteration in chromosome segregation of some mutant strains.[3,4] Nevertheless, alterations in sister chromatids division during mitosis can result from a vast range of factors. Between these, alterations that perturb the structure and/or the high dynamicity of microtubules are key factors able to impair chromosomes segregation, mainly because of the subsequent inability to form the mitotic spindle.[5] Given the central role that microtubules play in the MC process, it has to be pointed out that the term "catastrophe" associated with a phenotype of cellular death reported 5 years before in 1989, when McIntosh JR. proved that microtubule can undergo a process of disintegration that he called "microtubule catastrophe". [6] What is then mitotic catastrophe? According to what was demonstrated in references [3,4], some thought this process happens to mammalian cells unable to fulfill a complete mitotic process , which would result in tetraploidy, a double chromosome quantities (or 2N, after a single cell cycle),[7] or endopolyploidy,[8] a situation characterized by multiple chromosome settings in a single cell (or XN, several than a cell cycle). Actually, even if exact definitions of MC are still missing, MC is prevalently defined as a cell death path that is activated by premature or inappropriate entry of cells into mitosis, and that can be activated upon chemical or physical stresses.[9] Recent evidence demonstrates that MC is not a distinguished way of cell death, but rather represents a *"pre-stage"* anticipating cell death, which later occurs via necrosis or apoptosis.[9] Moreover, even though it is widely accepted that MC is the main outcome after ionizing radiation treatment or administration of drugs that influence microtubule assembly and stability, thus inducing mitotic failure, the final cell death pathway and the final outcome of MC (Figure 2) strongly depend on the cell type and its related molecular profile.[9,10] These molecular profiles still need to be fully elucidated and classified. However, some key features of MC are already established, as failure in DNA repair mechanisms, genomic instability, chromosome segregation impairment, and microtubules destabilization. [6,11] But how these factors can drive through MC? It is well established that to preserve genome integrity, DNA-damaged responses can either result in cell-cycle stalling, finalized to give time to activate DNA repair mechanisms, or in the removal of cells that are irreparably injured via apoptosis, MC, or necrosis. At the same time, drugs that influence microtubule stability, influencing mitotic spindle formation or chromosomes segregation, end up in the elimination of cells incapable to complete a correct mitosis. The same can be said about deficiency in cell-cycle checkpoints (especially the ones related to DNA structure analysis and to mitotic spindle): when these checkpoints are compromised, cells might prematurely enter into mitosis (M phase), as when DNA has still not been repaired, thus going toward MC.[9,12] MC-related alternations are associated with morphological changes, as the acquisition of big cellular dimensions and the formation of abnormal nuclei, mainly due to micronucleation (chromosomes or chromosome fragments not evenly segregated into the two daughter nuclei) and multinucleation (missegregated chromosome fragments enveloped by one or multiple nuclear membranes of different dimension). The latter two cases generally represent the final step of MC.[9] By this description, it appears evident that MC is a type of mitosis-related cell death that occurs during or shortly after a defective or failed mitosis.[3,13] This concept is further corroborated by the evidences that MC is also associated with incomplete DNA duplication and premature chromosomes condensation (PCC).[14–16] Nevertheless, chromatin condensation is a key feature of another well-defined type of cell death, apoptosis, which, anyway, can be distinguished from MC by the TUNEL (Terminal deoxynucleotidyl transferase dUTP Nick End Labeling) test, which individuates nuclear fragmentation, and by cell shrinkage. The morphological dissimilarities let, initially, hypothesize that MC was a direct cause of death, not connected with apoptosis, a hypothesis further corroborated by the evidence that inhibition of caspases did not influence the formation of giant cells.[10] Nevertheless, some papers suggest that MC could also be associated with caspase-2 and 3 activation, mitochondrial release of pro-apoptotic factors, and DNA fragmentation, all apoptotic features.[13,17,18] This initially created debate about MC: is it a distinct type of cell death, or is MC only a "pre-stage" of apoptosis? Additional questions arise when it was demonstrated that heated HeLa cells underwent PCC and premature reconstitution of nuclear envelope around asymmetric cluster, but did not ended up in nuclear fragmentation and apoptosis, but rather in the formation of multiple micronuclei that lead to a necrotic-like death.[14] Moreover, MC has also been characterized as delayed form of reproductive death, described as the inability to form a viable progeny able to undergo new mitotic divisions, since giant cells exhibiting multiple nuclei can be temporally viable.[19–21] This underlines that a big debate is still going on about MC, which, by some, is seen as distinct type of cell death, while others believe that it is a process that ultimately leads to apoptosis or necrosis. Agreement has been found on the concept that MC is a mitotic-related cell death and that, without entering the M phase, a cell cannot undergo MC.[10] Since MC is induced by DNA damage, as well as by alterations in cell-cycle checkpoints, it might be possible that MC is a defense mechanism to avoid aneuploidization, which might bring to neoplastic transformation. This is likely to be stimulated not only by DNA alterations, but also by deficiencies in various proteins involved in cell-cycle regulation, as cell-cycle-specific kinases Cdk1 (cyclin B1-dependent kinases) or Aurora kinases, cell-cycle checkpoint proteins and mitotic spindle assembly as Chk1 (checkpoint kinase 1) and Chk2 (checkpoint kinase 2), or cell death inducers as p53, p21 and caspases.[10] When the DNA damage is the inducer of MC, which can be induced by various chemotherapeutic agents, cells generally die during interphase, before entering the M phase. If they survive to interphase,

DNA-altered cells will stop in G1 or G2 to restore the damage and, eventually, reenter the cycle. Nevertheless, DNA alterations are not always repaired, leading to the activation of the pathways of ATM (ataxia-telangiectasia-mutated) kinases and of ATR (ataxia-telangiectasia and Rad3-related) kinases which, through an evolutionary conserved kinases cascade, activate the checkpoint-regulated Chk1 and Chk2.[22,23] Chk1 and Chk2 kinases are especially essential for cell-cycle arrest in G1, S, and G2 phases. Chk1 is phosphorylated, and activated, in response to DNA alterations, and seems to play a role at every point of the cell cycle.[24,25] Indeed, the conditional knockout of Chk1 in transgenic mice induces a vast cell death in proliferating somatic cells, which show clearly visible morphological alterations in the nucleus typical of cell that are undergoing MC.[26] To further demonstrate that an intact DNA structure checkpoint is an essential prerequisite to prevent MC, we can look at the substrate targets of the Chk2 kinases, such as p53,[27,28] the cell-cycle-regulating phosphatases Cdc25A and Cdc25C[29,30] and Mdm2,[28] which are not only required for MC, but appear to have an apoptosis-sensitizing effect,[13] remarking the interconnection between MC and apoptosis. Anyway, some authors point on the differences between MC and apoptosis and sustain that the two pathways are distinct,[31] since some manipulations able to block the apoptotic cascade (as overexpression of Bcl-2) increase the frequency of MC.[32] To further sustain the hypothesis that MC is unrelated to apoptosis, Nabha *et al.* demonstrated that caspase inhibitors such as Z-VAD.fmk are unable to halt the appearance of giant multinucleated cells, induced by treatment with spindle poisons.[33] However, the latter demonstration does not account for the possibility that apoptosis might not be mediated by caspases, which has been proved by different authors.[34–36] Moreover, the pharmacological inhibition or genetic deletion of genes related to the "DNA structure checkpoint" such as ATM, ATR, Chk1 and Chk2, lead to DNA damage-induced MC.[31,37–39] Thus, inhibition of this checkpoint limits the time available for DNA repair, constraining the cell to prematurely advance through G1, S and G2 phase and leading to MC,[40,41] a process which is markedly different from apoptosis, yet at macroscopic level, due to large cells formation characterized by multiple micronuclei clustered around individual or group of chromosomes.[31,42–45] Mitotic phase entry, however, relies essentially on cyclin B/Cdk1 kinase (or cdc2) action, which must be timely activated and form a complex that gain functional activity.[46–48] Cyclin B transcriptional levels and activity begin during late S phase and peak in M phase, while its active form translocates to the nucleus in early mitosis.[46,49] On the other side, Cdk1 is kept inactivated by the phosphorylation operated by Myt1 and Wee1 until early mitosis, when the phosphate groups are removed by a member of the Cdc25 family phosphatases and CAK kinases phosphorylate Cdk1, ensuring its maximal activity to enter in early mitosis.[47] The interactions between these two elements give rise to the mitosis-promoting factor (MPF), which ensures the G2 to M phase transition. The Cdk1/cyclin B1 heterodimer induces mitosis by activating phosphorylation of enzymes that control chromatin condensation, nuclear membrane disaggregation, and microtubule restructuration, leading to cellular roundup.[50] The perfect coordination of spatiotemporal patterns of Cdk1/cyclin B1 activity is crucial to ensure a regular cell cycle and is controlled at multiple steps. Thus, the so-called DNA structure checkpoints, by activating Wee1 and Myt1, as well as the checkpoint kinases Chk1 and Chk2, avoid Cdk1 action and entry into M phase. While the "DNA structure checkpoint" stops cells before entering the M phase in case of DNA alteration or not complete duplication, another checkpoint, the so-called spindle assembly checkpoint, prevents anaphase entry until all chromosomes are bound on both sides with the mitotic spindle. Indeed, this will ensure a balanced chromosome segregation between daughter cells.[47,51,52] To do this, the "spindle assembly checkpoint" promotes the formation of the anaphase-promoting complex (APC), which ensures the rapid destruction of cyclin B1 thanks to its E3 ubiquitin ligase activity.[50,53,54] Moreover, APC-mediated ubiquitination of securin allows the separase to bind securin itself, event that ends in the breakdown of the cohesins that maintain the linkage between sister chromatids. By this, sister chromatid segregation and anaphase entry are granted. When alteration in the mitotic spindle dynamics activates this checkpoint, the cell stops M phase before anaphase onset. This may lead to prolonged inhibition of APC (extended Cdk1 temporal activity), which is known to induce MC generally characterized by centrosome overduplication. Analogously, defects in checkpoint signaling are able to lead to abortive centrosome duplication,[55] multipolar mitosis, and premature segregation of unaligned chromosomes with uneven partition of genetic material into the offspring.[56,57] This causes MC. In this way, the "spindle assembly checkpoint" prevents aneuploidy by allowing unbounded kinetochores on chromosomes that would be missegregated to delay the transition from metaphase to anaphase until they become appropriately attached. This is why MC can be induced by drugs that, by acting on mitotic spindle formation and cell division, affect progression through mitosis. Indeed, spindle poisons are mitotic inhibitors that have various impacts on microtubule dynamics, since by binding with tubulin they can prevent either microtubules assembly or their disassembly. Thus, altered assemblage of mitotic spindle deficits in components of "spindle assembly checkpoint" and regulators of cell cycle, as DNA damage can be associated with MC. Moreover, it has to be remembered that MC induced during experimental procedures greatly depends on the genetic background, energy, and metabolism state of the cell.[9] An important concept when talking about MC is the one related to checkpoint adaptation and mitotic slippage. Checkpoint adaptation is the capacity of a cell to enter the M phase even after a prolonged checkpoint-imposed cell-cycle arrest in the presence, for instance, of DNA alterations.[58,59] These cells are committed to become arrested before entering anaphase, which is a way that conduce to MC as just described,[55] thus inducing to hypothesize that checkpoint adaptation might represent a different way to remove cells with unrecoverable damage.[60] Similar to adaptation, mitotic slippage promotes the inhibition of checkpoint activity by prolonging temporal extension of mitotic arrest. This leads to the development of tetraploidy in cells that can complete mitosis thanks to the constitutive activation and/or permanence of cyclin B1, which should be, instead, degraded by APC to allow progression to anaphase.[61–63] However, when the arrest is prolonged, cyclin B1 cannot escape from the slow but continuous proteasomal degradation, thus forcing cells to exit from M phase, with subsequent nuclear envelope reconstitution around cluster of misaggregated and altered chromosomes to form multiple nuclei.[62] These tetraploid cells will subsequently experience a definitive arrest in G1 state due to checkpoint activation. Mitotic slippage is one of the important characteristics of cell undergoing MC: they increase in size up to 200 times and become giant cells with anomalous nuclei.[20,64] The presence of multinucleated, MC-derived cells can be easily distinguished from apoptotic death by carefully observing the morphology of their nuclei. Apoptotic cells are, indeed, characterized by nuclear fragmentation, condensed chromatin, and cellular shrinkage; on the contrary, MC is identified by the formation of clusters of nuclear envelopes that contains misaggregated chromosomes with various and different grade of chromatin condensation.[9,14,41] But how do MC-related giant cells die? Notwithstanding the morphological differences between apoptosis and MC, they do share biochemical hallmarks, as further demonstration that MC represents a pre-stage of apoptosis. Indeed, Chk2 inhibition is able to induce MC in HeLa cells, a process that was associated with caspase-2 and caspase-3 activation, permeabilization of mitochondria that release apoptotic-involved factors (as cytochrome c) and DNA fragmentation.[14,16, 65] Moreover, caspases appear to be essential in the final stages of MC, thus supporting the hypothesis that morphological alterations associated with MC are just the initial steps that will lead to apoptosis.[12,66] To further sustain this concept, it has been proved that mitotic slippage and mitotic arrest address to apoptotic death.[67–69] Nevertheless, MC-mediated death do not always require caspases activation and apoptosis, as demonstrated by the capacity of some giant cell to undergo slow death in a necrosis-like fashion, with loss of nuclear and plasma membrane integrities ending in cell lysis.[13] Indeed, some evidences demonstrated that MC can happen in a caspase-independent fashion, since MC can induce several rounds of abnormal mitosis, associated with polyploidy plus multinucleation, which direct the cells to necrotic death.[70] However, a central point when talking about MC is p53. Several studies proved that spindle-damaged cells that are able to cross the M phase will be halted, in a tetraploid state, during G1 phase thanks to the activation of G1 checkpoint by p53.[71,72] This, in its turns, irreversibly activates p21, thus blocking cell-cycle progression of tetraploid cells in order to avoid altered-cell propagation. This will all end into cell death.[7,71] Moreover, p53 is the key factor that activates apoptosis after endomitosis and endoreplication. p53 and apoptosis, anyway, are not always required for MC-related cell death. Several papers evidenced how MC and necrosis can be morphologically distinguished especially when referring to loss of nuclear membrane and plasma membrane integrities. [42,73] While more detailed characteristics of MC still remain elusive and need to be investigated to elucidate all the different involved pathways, great consensus can be found on the determinants that will push MC-associated cell to choose between apoptosis or even necrosis: the genetic status of the cell and the profile of proteins involved in cell-cycle regulation.[13]

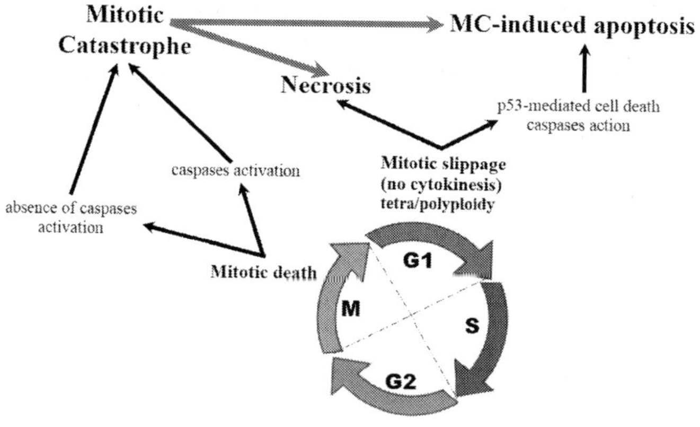

Figure 2. Schematic representation of main cell death path related to MC.

4. Cell-cycle alterations in post-mitotic neurons

Mature neurons of central nervous system are typically described as permanently post-mitotic cells that have been recently revealed to be in a continuously activated but arrested cell-cycle status. Thus, neurons must constantly keep their cell cycle in check, avoiding its re-initiation, since vigilance relaxation would mean death. During the development, neurogenesis takes place mainly in a tightly packed layer of nuclei lining the lumen of the neural tube (the ventricular zone, VZ) and later in the closely apposed region known as the subventricular zone (SVZ). After birth, the VZ has been depleted of all mitotic cells, but cells with stem cell precursors properties are present throughout the adult brain particularly in the SVZ (to a greater or lesser extent in different vertebrates) and can give rise to neurons in the adult in case of injury. Following the last cell division, neurons mature and the still unclarified mechanisms that will ensure a permanent mitotic arrest begin. What would happen if a neuron lost control of its cell cycle and reentered cell division? Evidence suggests that the neuron would die. One of the first descriptions of this phenomenon was given in 1992 analyzing the effects of expression of SV40 T antigen in Purkinje cells or in photoreceptor. While the expression of this viral oncogene typically promotes tumorigenesis in mammalian cells, its expression in post-mitotic neurons induced the appearance of mitotic figures, and entry in S phase not followed by proliferation but by cell death.[74,75] RB protein has an important role in the maintenance of neuron cell-cycle control and mice deficient in Rb show in the nervous system, ectopic mitoses, and massive neuronal death.[76] Freeman *et al.* (1994)[77] demonstrated that the death of cultured sympathetic neurons following NGF deprivation led to the upregulation of cyclin D1. These experimental data support the hypothesis of an association between ectopic cell division and cell death in the nervous system. All these data converge on the idea that during development, once a neuron leaves the VZ, its cell cycle must be actively held in check and cell-cycle re-initiation leads to death. In the adult life, cell-cycle re-activation has been claimed as the cause of neuronal death in many neurologic disorders such as Alzheimer's disease (AD), amyotrophic lateral sclerosis (ASL), ataxia telangiectasia, Parkinson's disease (PD), and even stroke.[78] In the early nineties of the last century, it became increasingly clear that mitogenic pathways in neurons are aberrantly activated early during AD. In the neurons of patients who had died with AD have been reported the presence of phosphorylated tau protein, usually found only in dividing cells[76] and of various cell-cycle proteins, such as cyclin A, B, D and E,[79,80] PCNA and CKIs of both the Ink and Cip/Kip families.[81] Evidence of cell-cycle activation in post-mitotic neurons has been also observed in murine models of AD, such as increased expression of the cell proliferation marker PCNA, and the CDK inhibitor Cdkn2a, as detected by immunohistochemistry in cortical neurons of the APP/PS1 mice.[82] Bonda *et al.* demonstrated that phosphorylated MCM2 protein, a component of the DNA replication machinery, is localized to the cytoplasm, rather than nucleus, in neurons of AD patients, while it was completely absent in control cases.[83] These data support the notion that neurons in AD reenter the cell cycle and progress through the S phase, but the aberrant cellular distribution of MCM2 may induce cell-cycle stasis and consequent neuronal degeneration. Entry into the S phase not followed by progression through the M phase in dying neurons in AD patients has been reported also by others.[84] Similarly, Ogawa *et al.*, using immunohistochemistry, reported an increase in phosphorylated histone H3 in hippocampal neurons in AD patients. However, phosphorylated histone H3 is localized within the nucleus in active dividing cells, while in neurons of AD patients the proteins were aberrantly localized in the cell cytoplasm, indicating that cells underwent MC that leads to neuronal dysfunction and neurodegeneration. [85] In animals, AD-like neurodegeneration can be experimentally induced by the administration of okadaic acid (OA), a potent phosphatase inhibitor[86,87] that increases the degree of phosphorylation of various proteins, such as the microtubule-associated protein tau. *In vitro* exposure of neuroblastoma cells to OA induces the expression of G2/M phase marker cyclin B1 and cyclin D1,[88] causing neurons to aberrantly reenter cell cycle, but the cells fail to complete the mitotic cycle, resulting in MC. Indeed, Chen *et al.* demonstrated that disturbance of the protein kinase–phosphatase system caused by OA is sufficient to induce neuronal cyclin B1 expression, force neurons into the mitotic phase of cell cycle, and cause MC.[89] Although the presence of binucleated neurons has been reported,[90] no cytokinesis has ever been described, consistent with the idea that susceptible neurons may be arrested at the G2/M transition before they die. In mitotic-competent cells, G2 phase is associated with the activation of CDK1 and phosphorylation of target proteins, that is, microtubule-associated protein tau, that allow cell-cycle progression.[91,92] However, in neuron the presence of a highly specialized and stabilized cytoskeleton probably hinders the possibility to disassemble cytoskeleton structures to commit to mitotic spindle formation and cytokinesis. Interestingly, in AD patients the presence of abnormally increased level of phosphorylated tau has been reported and this could be explained as an unsuccessful effort to modulate G2 neuronal architecture and prepare it for mitosis.[93] In addition, the pro-apoptotic BAD protein is another protein phosphorylated and activated by CDK1 in G2 phase, linking the cell cycle to the programmed cell death mechanisms.[94] However, the apoptotic nature of neuron death in AD remains controversial. Indeed, although extensive DNA fragmentation has been reported, only a small number of cells show signs of caspase-mediated apoptosis. Interestingly, activation of caspases in human neurons does not lead to a rapid process of cell death but provokes a prolonged form of apoptosis, thus explaining the long-term survival of neurons that have progressed in late phases of the cell cycle and the protracted nature of AD. Indeed, quantification of neuron-expressing cell-cycle proteins revealed that 5-10% of neurons are dying at any moment. This implies that death by cell cycle in adult neurons must be a very slow process requiring in the order of 6–12 months; otherwise, during the average course of AD (10 years), 95% of neurons should be dead in less than a month. Thus, it is possible that in fully mature neurons, neuronal death by cell-cycle re-initiation requires an additional stimulus to make the transition from cycle to death.

The involvement of cell-cycle molecules was found also in ALS. Nuclear accumulation of phosphorylated RB protein, with concurrent increase in cytoplasmic levels of cyclin D, and redistribution of E2F-1 into the cytoplasm occur in motor neurons and glia during ALS, suggesting overcoming of G1/S checkpoint during ALS as mechanisms regulating motor neuron death.[95] More recently, Ranganathan *et al.* linked elevated levels of p53 in ALS spinal cord motor neurons and activation of G1 to S phase cell-cycle regulators to their cell death.[96] Using microarray analysis in single motor neuron in the SOD1 transgenic murine model of familial ALS, researchers found a significant increase in the expression of cyclins D2, E2, and

I.[97] In ataxia-telangiectasia (A-T), the extent of Purkinje cell loss correlates with the severity of the clinical phenotype. Re-expression of cell-cycle proteins in the cerebellum of patients has been reported in particular in Purkinje cells and striatal neurons in both human and mouse A-T.[98] This linkage of the neurodegeneration with events of the cell cycle has important implications in understanding the etiology of A-T. The involvement of cell-cycle-associated mechanisms to neurodegeneration in PD is suggested by a series of experimental evidence. In human, the expression of phosphorylated Rb protein, E2F1, as well as DNA synthesis, has been reported in postmortem samples of PD patients.[99,100] Administration of 6-hydroxydopamine (6-OHDA) induced a rat model of PD in which overexpression of Cdc2 was found in dopaminergic neurons of the substantia nigra.[101] *In vitro* treatment of PC12 cells with 6-OHDA lead to oxidative stress and was associated with upregulated expression of several markers of cell-cycle reactivation.[102] The role of cell-cycle reentry in neuronal dysfunction and death observed in many neurodegenerative diseases has been clarified by Lee *et al.*.[103] These researchers developed a new transgenic mouse model in which forebrain neurons were induced to reenter the cell cycle by overexpression of the proto-oncogene Myc under the control of the CAMKII promoter. In this animal model, the induction of ectopic cell-cycle reentry (determined by the expression of PCNA, Ki-67 and cyclin D1, and BrdU incorporation) results in neuronal cell death, gliosis, and cognitive deficit, thus conclusively demonstrating that dysregulation of cell-cycle reentry is a key determinant of neurodegeneration *in vivo*. Activation of the cell cycle mediated by the loss of CDKI function is also reported as a cause of ischemia-induced delayed neuronal death *in vivo* and *in vitro*.[104] Similarly, the phosphorylation of the RB protein is altered by transient brain ischemia.[105] However, the activation of cell cycle and the rapid expression of cell-cycle inhibitors, such as p53 and p21cip1, may trigger neuronal death in acute hypoxic stress,[106] while preventing cell death in the presence of sublethal stimulus.[107] All these data suggest that cell cycle is aborted mostly at G1/S checkpoint in ischemic death. However, in some systems, authors have shown that neurons start to replicate DNA, thus entering the S phase before they die, as demonstrated by induction of PCNA and incorporation of BrDU in several models of focal and global ischemia.[108–110] However, there is no *in vivo* evidence of G2 entry in ischemic neurons. In spite of the accumulating *in vivo* experimental studies demonstrating aberrant deregulation of cell cycle in stroke, there is little known about how/whether these processes occur in humans. Studies in human biopsy material indicate that the neuronal death in temporal lobe epilepsy is also associated with the expression of cell-cycle-related proteins.[111] Finally, in adult central nervous system, some DNA-damaging agents, such as UV radiation and cytotoxic drugs, have deleterious effects because they induce neuronal death via the activation of the cell division cycle[112] and the cell death follows a MC that can be prevented by cyclin-dependent kinase inhibitors.[112,113] The evidence to date is compatible with the requirement for a second "hit" for a neuron to progress cell-cycle initiation and DNA replication to death. This finding offers several important insights. The most important is that any intervention of blocking "second" processes might prevent or slow the neuronal cell death in the process of disease. Finally, which type of cell death follows the reported cell-cycle alterations in post-mitotic neurons is not completely understood, but MC is one possibility.

5. Cell-cycle alterations in podocytes

Podocytes are specialized renal epithelial cells that, through cytoplasmic extensions called foot processes, interdigitate with neighboring podocytes and cover the surface of the glomerular capillary loops, thus forming the glomerular filtration barrier (GBM). Podocytes, like neurons, are terminally differentiated post-mitotic cells, with a sophisticated actin cytoskeleton, whose disruption due to genetic, mechanic, immunologic, or toxic injury leads to the detachment of cells from glomerular basement membrane, and finally to podocyte loss. Decrease in the number of podocytes in the glomerular capillary tuft is associated with the development of glomerular sclerosis in several human and experimental diseases. Two cellular strategies that could act to compensate for cell stress or relative cytopenia (e.g., during organogenesis or injury) are hypertrophy, which is an increase in cell size, and hyperplasia, which is an increase in cell number. Both these processes require that quiescent cells reenter the G1 phase to increase the amount of cell organelles and proteins. Podocytes can only undergo hypertrophy, producing additional foot processes to compensate for podocytopenia. Indeed, several protective mechanisms prevent podocyte progression to mitosis and arrest their cell cycle at the restriction points of G1 and G2 phases. The ability of podocytes to arrest at the restriction point is well documented. *In vitro* exposure of quiescent mesangial cells and podocytes to antibody and a complement source induces sublytic injury and entry into G1 phase.[114,115] Mesangial cells progress through the cell cycle, synthesized DNA, and divide, with resultant proliferation and increased cell number. In contrast, podocytes undergo limited DNA synthesis, but do not proceed beyond the G2/M phase of the cycle and do not proliferate.[116] This arrest is associated with an increase in protein levels for p53, the CDKI p21, growth-arrest DNA damage-45 (GADD45), and the checkpoint kinases 1 and 2.[116] Sublytic C5b-9 injury causes DNA damage in podocytes, but not in mesangial cells,[117] and DNA damage prevents proliferation by arresting cells at G2/M phase. Re-expression of cell-cycle proteins has been reported during glomerular disorders, and in recent years, cell-cycle regulatory proteins have become an area of intense research in order to understand the changes that occur in various renal diseases. The roles of these proteins in renal diseases have been addressed in several reviews:[118–120] cyclin A and Ki-67 staining was observed in podocytes of children collapsing glomerulopathy[121,122] and focal segmental glomerulosclerosis (FSGS);[123] cyclin D was observed in cellular lesion of FSGS.[123] Moreover, p27 and p21 but not p57 was decreased in CGN (crescentic glomerulonephritis), as in FSGS when compared to normal.[122] A uniform decrease in p27 and p57 immunostaining in FSGS and collapsing glomerulopathy was reported also by Shankland *et al.*.[124] However, these authors reported de novo expression of p21. Downregulation in the expression of cyclin kinase inhibitors such as p21 and p27 has been observed in podocytes in children with idiopathic nephrotic syndrome. This downregulation is not followed by upregulation of cyclin D and cyclin A that are needed to overcome the G1/S transition and move the cell forward in the cell-cycle process. Thus, the podocytes remain trapped in the G1 arrest phase. Similarly, an altered expression of 27, p21, and cyclin A was reported by Srivastava and colleagues also in patients with minimal change disease,[122] while Shankland *et al.* did not reported change in expression of p21, p27 and 57 in podocytes of patients affected by minimal change disease and membranous glomerulopathy.[124] Using animal models of glomerular disease such as the passive Heymann nephritis (PHN) and antiglomerular antibody models, Hiromura *et al.* reported a marked decrease in p57 expression that was diffuse in PHN, whereas in the murine model, loss of expression of p57 occurred predominantly in podocytes expressing PCNA.[125] Marked podocyte expression of the CKIs, such as p21 and p27, was reported during Heymann nephritis in diabetic ZDF-fa/fa rats,[126, 127] and in glomerular tufts affected by crescentic glomerulonephritis,[128] suggesting upregulation of CKIs as a generalized response of podocytes to stress or injury. Interestingly, mechanical stretch of cultured podocytes reduced cell-cycle progression and induced hypertrophy in both wild-type and p27−/− podocytes.[129] Moreover, this event required the presence of p21 and was prevented by specifically blocking extracellular signal-regulated kinase 1/2 (Erk1/2) or Akt.[129] These data suggest that upregulation of CKIs in podocytes is an attempt to maintain cell-cycle quiescence and preserve normal physiological function. However, it is possible to overcome the resistance and force podocytes entering the cell cycle, under the pressure of sufficiently strong stimuli. We recently provided evidence that the dramatic consequences of this forced entry in mitosis are the trigger for a catastrophic mitosis. Indeed, as already reported for neurons, podocytes are programmed to maintain their complex cytoskeleton and cannot assemble an efficient mitotic spindle due to poor expression of Aurora kinase B, which is essential for cytokinesis[130] (Figure 3). Thus, podocytes cannot survive mitosis and, if forced to override these cell-cycle restriction points, detach and are lost in the urine. This is confirmed by the presence of many binucleate podocytes in the urine in patients affected by FSGS[131] and lupus nephritis,[132] suggesting that podocytes carrying nuclear abnormalities generated during an abnormal cytokinesis are more susceptible to detachment and loss. Moreover, in mitotic podocytes actin is required for the formation of the mitotic spindle and is no more compatible with maintaining the cytoskeletal structure of secondary foot processes. Indeed, in adherent cells the actin network is rapidly dismantled and rearranged to allow the cell to form the mitotic spindle and to enter mitosis. Therefore, mitotic cells acquire a rather rounded shape and easily detached and are lost in urine. Thus, acquisition of functional specialization in a cell type, such as podocytes, neurons, and cardiomyocytes, is coupled with the permanent exit from the cell cycle[133] and the arrest in a "post-mitotic" state. Therefore, when expression of cell-cycle markers (such as Ki-67) in podocytes is observed in certain diseases this is probably a sign that they undergo hypertrophy, but they are unlikely to undergo mitosis. Podocytes with mitotic figures are only occasionally observed because they are susceptible to detachment and/or death. Nagata *et al.* reported one mitotic figure in a single podocyte in a case of FSGS among 164 renal biopsy specimens with glomerular disease.[134] Mitotic podocytes associated with proteinuria may be a desperate but aborted attempt to regenerate epithelial injury.

Podocyte multinucleation on the other hand is a recognized feature of aberrant mitosis, also described by Nagata *et al.* in 1998.[135] More recently, Mulay *et al.* in lieu of the fact that aberrant mitosis and podocyte binucleation are synonymous to MC, retrospectively reviewed the consecutive renal biopsy specimens and found twice as many (n=12) multinucleated podocytes.[136] All cases had significant proteinuria, and diagnoses included minimal change, FSGS, IgA nephropathy, membranous nephropathy, collapsing glomerulopathy, and membranoproliferative glomerulonephritis. Foot process effacement was invariably present in association with binucleated podocytes. These binucleated, and sometimes trinucleated, podocytes enlarge, with edematous nuclei, with one or more

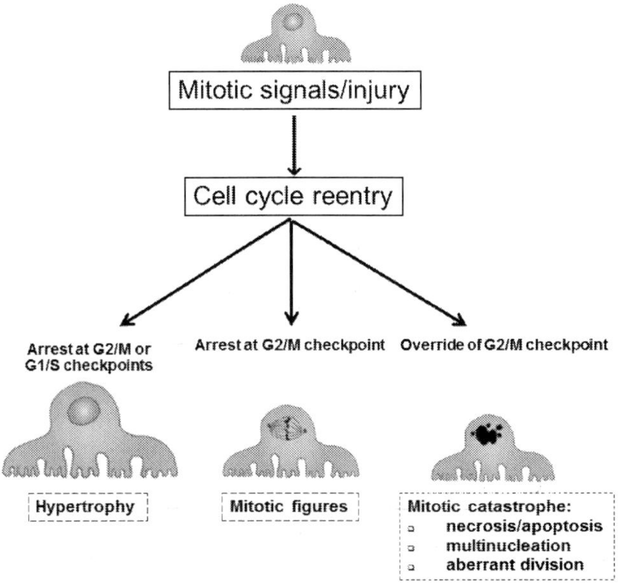

Figure 3. Podocytes are terminally differentiated post-mitotic cells. Pathological activation such as exposure to stress induces a compensatory response. The most important response to stress is podocyte hypertrophy, which results from cell-cycle arrest at the G1 or G2/M checkpoint. Persistence of mitogenic stimuli forces the podocyte to complete mitosis. This could lead to the formation of mitotic spindles and aberrant divisions with the formation of aneuploid cells or to the execution of a mitotic death program.

nucleoli. The presence of micronuclei is also reported and it is indicative of genomic instability and characteristic of MC. Moreover, these podocytes exhibit other alteration typical of MC, such as increased number of organelles, mitochondria and lipid droplets or cytoplasmic vacuoles.[137] The classic example of mature podocytes reentering the cell cycle is HIV-associated nephropathy (HIVAN). In HIVAN, the presence of tightly packed multinucleated podocyte is frequently observed.[138,139] These multinucleated podocytes to large extent remain attached to the GBM, and there is no evidence of overt foot process effacement and of apoptotic nuclear condensation. The cytoplasm appears fragile with cytoplasmic dense (osmiophilic) bodies (lysosomes). Eventually, these multinucleated podocytes detach from the GBM, disrupt the cytoplasm, and release the nucleus and cytoplasmic contents into the Bowman space.[137] Podocyte mitoses can be seen.[137] However, the connection with MC and aberrant cell death is still vague and little studied experimentally or clinically. Interestingly, in HIVAN p27, p57, and cyclin D are absent in podocytes, and p21, cyclin A, and Ki-67 are induced.[121,124,140] This implies a dysregulated podocyte phenotype characterized by bypassing cell-cycle restriction points and podocyte loss via MC. Recently, we demonstrated that Notch activation may represent an important driver of MC in podocytes during glomerular disorders.[130] Indeed, Notch activation in podocytes *in vitro* induced the downregulation of the p21 and p27, which pushed progression toward mitosis of a cell that cannot assemble an efficient mitotic spindle, leading to the formation of bi- or micro-nucleated cells with disrupted cytoskeleton.[130] Accordingly, Notch protein expression was not detected in glomeruli of healthy adult kidneys, while several studies demonstrated strong Notch upregulation in podocytes of patients affected by several types of glomerular disorders characterized by podocyte death.[130,141,142]

6. Cell-cycle alterations in renal tubular epithelial cells

6.1. G1 arrest

Acute kidney injury (AKI) is a potentially devastating, increasingly common syndrome characterized by rapid impairment of kidney function as a result of a toxic or ischemic insult. In the first 24 h following injury, tubular cells undergo apoptotic and necrotic cell death, and 70% of the surviving, normally quiescent proximal tubule epithelial cells enter the S phase of the cell cycle.[143,144] This is documented by an increase in PCNA,[145–147] incorporation of 3H-thymidine and 5-bromo-2-deoxyuridine into nuclear DNA, and induction of mRNA for "immediate-early" genes, c-fos, c-jun, and egr-1.[148,149] A rapid induction of p21, in several models of AKI, has also been reported.[149] This cell-cycle reentry after injury has traditionally been viewed as an appropriate repair response to the loss of adjacent cells after an initial insult. However, this is in contrast to the observation that cell-cycle inhibition is protective against several form of AKI. The strategic role of p21 in AKI is demonstrated by a series of experiments. Administration of an adenovirus vector directing the expression of p21 protects mouse proximal tubule cells in culture from cisplatin toxicity.[150] Similar results can be obtained by treating cells with several cell-cycle inhibitors, such as roscovitine and olomoucine.[150] *In vivo*, Megyesi *et al.*[151] demonstrated the upregulation of p21 mRNA in ARF (Acute Renal Failure) induced by ischemia, cisplatin, or ureteral obstruction. In addition, p21 knockout mice were shown to have increased kidney cell-cycle activity, increased cisplatin nephrotoxicity, and higher mortality than wild-type animals.[145] In p21 (−/−) mice, 5-bromo-2-deoxyuridine incorporation into nuclear DNA and increases in PCNA content were much higher compared with p21(+/+) mice. Following either cisplatin administration or ischemia–reperfusion injury (IRI), compared with their p21(+/+) littermates, p21(−/−) mice developed more severe morphological damage, displayed a more rapid onset of the physiological signs of AKI, and had a higher mortality.[152] These findings suggested that the induction of p21 plays a protective role in kidney cells by preventing DNA-damaged cells from progressing in the cell cycle without repair, which eventually would result in death. Miyaji *et al.*[153] speculated that p21 induction may play a protective role to prevent the development of cisplatin-induced AKI, stopping the cell cycle in the G1 phase and providing enough time for DNA repair. Similarly, Nath and co-workers[154] reported that LLC-PK1 kidney cells overexpressing heme oxygenase-1 were resistant to several apoptotic stimuli and that this resistance was mediated by p21 upregulation. Similarly, Golzalez-Michaca *et al.*[155] demonstrated in a rat proximal tubule cell culture that blocking the apoptotic effects of hemin was correlated with upregulation of p21 levels.

The hypothesis that cell-cycle inhibition post-insult protects against AKI is supported by several experimental findings. As reported above, the cell-cycle-inhibitory drug roscovitine is effective in protecting kidney cells *in vitro* from cisplatin-induced apoptosis.[150] Similarly, the broad spectrum small-molecule CDKI, purvalanol, can actually protect against cisplatin-induced cell death[156] and expression of an inactive Cdk2 mutant in mouse kidney cells protected from cisplatin nephropathy.[157] Importantly, small-molecule inhibition of CDK2 leads to a G2/M block or an intra-S phase arrest, and G2/M arrest would likely be an undesirable long-term effect as it has been recently reported that this induces progressive interstitial fibrosis in the kidney and increased cell apoptosis (see following paragraph). Interestingly, early G1 and late S phases are reported to be more resistant to genotoxic damage.[158] Thus, delaying this cell-cycle progression after injury should also ensure more time for DNA damage to be repaired before mitosis. DiRocco *et al.* recently demonstrated that a small-molecule inhibitor of CDK4/6 effectively promotes transient G0/G1 arrest in renal epithelial cells, protects these cells from DNA damage and apoptosis as a result of exposure to cytotoxic chemotherapeutic agents *in vitro,* and ameliorates kidney damage following AKI *in vivo*.[159] Their results suggest the possibility that epithelial cells may die through MC after IRI.

Thus, following injury, tubular cells enter the cell cycle, but rapidly arrest in the G1 phase. This G1 cell-cycle arrest prevents cells from dividing when the DNA may be damaged and arrests the process of cell division until the damage can be repaired lest resulting in the cell's demise or senescence. Two inducers of G1 cell-cycle arrest, insulin-like growth factor-binding protein 7 (IGFBP7) and tissue inhibitor of metalloproteinases-2 (TIMP-2), have been recently identified and their urinary levels serve as sensitive and specific biomarkers to early prediction of AKI and of renal recovery.[160,161] IGFBP7 directly increases the expression of p53 and p21 and TIMP-2 stimulates p27 expression. The upregulated p proteins in turn block cell-cycle promotion acting on the cyclin-dependent protein kinase complexes (CyclD-CDK4 and CyclE-CDK2), thereby inducing G1 cell cycle presumably to avoid cells with possible damage from dividing. Markers of cell-cycle arrest such as TIMP-2 and IGFBP7 may signal that the renal epithelium has been stressed and has shut down various function but may still be able to recover without permanent injury to the organ.[161]

6.2. G2/M arrest

For several decades, AKI was usually assumed to be transient with usual expected recovery of renal function if the individual survived the acute illness. Observational clinical studies and animal models however link AKI to chronic kidney disease (CKD) progression. When kidney injury is of mild entity with normal baseline function, the repair process can be adaptive with few long-term consequences. On the contrary, if injury is more severe, repeated, or to a kidney with underlying disease, the repair can be maladaptive. Maladaptive repair leads to CKD, a process characterized by persistent parenchymal inflammation, with increased numbers of myofibroblasts and accumulation of extracellular matrix (Figure 4). The mechanism that triggers the fibrogenic response after injury is not well understood but a G2/M arrest of tubular cells has been demonstrated being an important driver of maladaptive repair and progressive CKD after AKI.[162–164] Indeed, characterization of the cell-cycle profile of tubular epithelial cells *in vivo* at various times after an acute insult in multiple experimental models of murine kidney injury, including severe bilateral IRI, unilateral IRI, aristolochic acid-induced nephr-

http://dx.doi.org/10.5772/61783

opathy, and unilateral ureteral obstruction (UUO), identified the accumulation of cells in G2/M growth arrest as the common feature predicting progressive fibrotic kidney disease.[162] In the UUO model, authors demonstrated that fibrotic injury induced TGF-β1 secretion by tubular epithelial cells. TGF-β1 then induced G2/M cell-cycle arrest and profibrotic phenotype through the upregulation of p21 and activation of the JNK pathway, respectively. TGF-β1 and PDGF subsequently stimulated pericyte–myofibroblast transition through differentiation and proliferation, respectively.[165] In another study, authors demonstrated how severe IRI induces a sustained epidermal growth factor receptor (EGFR) activation, which is essential for the tubular cell regenerative response at the early stage of reperfusion, but eventually leads to the activation/proliferation of renal interstitial myofibroblasts and development of renal fibrosis.[164] The correlation between G2/M-arrest and fibrosis is supported by several studies. Pharmacological inhibition of G2/M-arrested cells reduced fibrosis, whereas increases in the proportion of G2/M-arrested cells in the cell-cycle exacerbated fibrosis, thus confirming that G2/M arrest in tubular cells contribute to progressive CKD after AKI.[164–166] The link between inadequate DNA repair and renal fibrosis and CKD is underlined in study on humans with the FAN1 mutation (a DNA damage response signaling pathway devoted to repair of DNA interstrand crosslink damage). These patients develop karyomegalic interstitial nephritis, with evidence of increased levels of DNA damage and cell-cycle arrest in the late G2 phase.[167] In conclusion, all these recent findings on the pathophysiology of AKI underline the important role of tubular cell-cycle arrest in the process of maladaptive repair. Moreover, these findings open new therapeutic perspectives to prevent, slow down, or arrest chronic fibrosis progression and progressive CKD, as suggested by experiments in rodents performed using agents such as histone deacetylase inhibitors[166] or p53 inhibitors[162,168], by blocking the initiation of the G2/M checkpoint, or by stimulating transit through G2/M to complete mitosis.

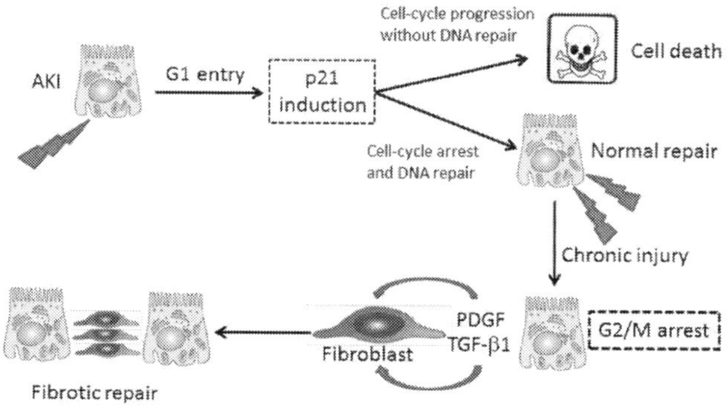

Figure 4. After AKI, normally quiescent proximal tubule epithelial cells enter the cell cycle, event traditionally viewed as an appropriate repair response to the loss of adjacent cells after an initial insult. However, rapid induction of p21 and arrest of cell-cycle progression play a protective role in kidney cells by preventing DNA-damaged cells from progressing in the cell cycle without repair, which eventually would result in death, probably by MC. In cases where the kidney injury is persistent or repeated, increasing numbers of epithelial cells stall between the G2 and M phases of the cell cycle, which stimulates the production of considerable amounts of TGF-β1 and PDGF that induce a fibrotic response. Pharmacological inhibition of G2/M-arrested cells reduced fibrosis, whereas increases in the proportion of G2/M-arrested cells in the cell-cycle exacerbated fibrosis, thus confirming that G2/M arrest in tubular cells contribute to progressive CKD after AKI.

However, application of these therapeutic strategies in humans needs careful assessment of the safety of the drugs and pathways under investigation because the G2/M checkpoint is extremely important in preventing the perpetuation of dangerous DNA mutations.

A great comprehension of the mechanisms involved in cell-cycle arrest could thus help not only in the discovery of novel therapeutic strategies to prevent podocyte loss, glomerulosclerosis, proteinuria and progressive kidney disease, but also in the selection and utilization of new specific and sensitive biomarkers for AKI.

7. Conclusions

Terminal differentiation invariably involves two closely linked phenomena: permanent withdrawal from the cell cycle and cell type-specific differentiation characterized by the upregulation of a panel of tissue-specific genes. Typically, post-mitotic cells do not reenter the cell cycle when exposed to growth signals, and in some cases further increases in tissue mass are achieved through an increase in cell size or hypertrophy. One long-standing theory to explain the lack of cytokinesis in post-mitotic cells, such as neurons, podocytes, and adult cardiac myocytes, is the presence of highly organized mature myofibrils which physically prevent cell division. Because cells must disassemble their cytoskeletal filaments before entering cell division, disassembly of the cytoarchitecture in these cell types would presumably negatively impact their function. However, in these cells expression of a wide range of cell-cycle proteins has been described, although no cases of cell division have ever been reported. This, together with the finding that the expression of cell-cycle proteins is necessary to execute cell death in response to certain stress signals, has led to the proposition that in post-mitotic cells, cell cycle is part of a well-regulated response to stress signals. The mechanisms by which cell-cycle reentry causes cell death are not completely known, but exploring the trigger(s) that induce normally post-mitotic cells to re-express cell-cycle proteins late in life as well as the molecular mechanism by which the induction of these proteins leads to cell death may produce great advances in the treatment and prevention of several neuro- and renal degenerative diseases.

8. Abbreviations

2N double chromosome quantity

3N triple to on chromosome quantity

MC mitotic catastrophe

MPF mitosis-promoting factor

M phase mitosis

PCC premature chromosomes condensation

http://dx.doi.org/10.5772/61783

Author details

Duccio Lombardi[1] and Laura Lasagni[1,2*]

*Address all correspondence to: laura.lasagni@unifi

1 Excellence Centre for Research, Transfer and High Education for the development of DE NOVO Therapies (DENOTHE), University of Florence, Florence, Italy

2 Department of Clinical and Experimental Medicine, University of Florence, Florence, Italy

References

[1] Murray AW. Recycling the cell cycle: cyclins revisited. Cell. 2004; 116:221–234.

[2] Arendt T. Synaptic plasticity and cell cycle activation in neurons are alternative effector pathways: the 'Dr. Jekyll and Mr. Hyde concept' of Alzheimer's disease or the yin and yang of neuroplasticity. Prog Neurobiol. 2003; 71(2–3):83–248.

[3] Molz L, Booher R, Young P, Beach D. cdc2 and the regulation of mitosis: six interacting mcs genes. Genetics. 1989; 122(4):773–782.

[4] Ayscough K, Hayles J, MacNeill SA, Nurse P. Cold-sensitive mutants of p34cdc2 that suppress a mitotic catastrophe phenotype in fission yeast. Mol Gen Genet. 1992; 232(3):344–350.

[5] Karsenti E, Newport J, Kirschner M. Respective roles of centrosomes and chromatin in the conversion of microtubule arrays from interphase to metaphase. J Cell Biol. 1984; 99(1 Pt 2):47s–54s.

[6] McIntosh JR. Cell biology. Microtubule catastrophe. Nature. 1984; 312(5991):196–197.

[7] Andreassen PR, Lacroix FB, Lohez OD, Margolis RL. Neither p21WAF1 nor 14-3-3sigma prevents G2 progression to mitotic catastrophe in human colon carcinoma cells after DNA damage, but p21WAF1 induces stable G1 arrest in resulting tetraploid cells. Cancer Res. 2001 Oct 15; 61(20):7660–7668.

[8] Ivanov A, Cragg MS, Erenpreisa J, Emzinsh D, Lukman H, Illidge TM. Endopolyploid cells produced after severe genotoxic damage have the potential to repair DNA double strand breaks. J Cell Sci. 2003; 116(Pt 20):4095–4106.

[9] Vakifahmetoglu H, Olsson M, Zhivotovsky B. Death through a tragedy: mitotic catastrophe. Cell Death Differ. 2008; 15(7):1153–1162.

[10] Surova O, Zhivotovsky B. Various modes of cell death induced by DNA damage. Oncogene. 2013; 32(33):3789–3797.

[11] Russell P, Nurse P. cdc25+ functions as an inducer in the mitotic control of fission yeast. Cell. 1986; 45(1):145–153.

[12] Vakifahmetoglu H, Olsson M, Tamm C, Heidari N, Orrenius S, Zhivotovsky B. DNA damage induces two distinct modes of cell death in ovarian carcinomas. Cell Death Differ. 2008; 15(3):555–566.

[13] Castedo M, Perfettini JL, Roumier T, Andreau K, Medema R, Kroemer G. Cell death by mitotic catastrophe: a molecular definition. Oncogene. 2004; 23(16):2825–2837.

[14] Swanson PE, Carroll SB, Zhang XF, Mackey MA. Spontaneous premature chromosome condensation, micronucleus formation, and non-apoptotic cell death in heated HeLa S3 cells. Ultrastructural observations. Am J Pathol. 1995; 146(4):963–971.

[15] Mackey MA, Morgan WF, Dewey WC. Nuclear fragmentation and premature chromosome condensation induced by heat shock in S-phase Chinese hamster ovary cells. Cancer Res. 1988; 48(22):6478–6483.

[16] Ianzini F, Mackey MA. Spontaneous premature chromosome condensation and mitotic catastrophe following irradiation of HeLa S3 cells. Int J Radiat Biol. 1997; 72(4): 409–421.

[17] Bataller M, Portugal J. Apoptosis and cell recovery in response to oxidative stress in p53-deficient prostate carcinoma cells. Arch Biochem Biophys. 2005; 437(2):151–158.

[18] Heald R, McLoughlin M, McKeon F. Human wee1 maintains mitotic timing by protecting the nucleus from cytoplasmically activated Cdc2 kinase. Cell. 1993; 74(3):463–474.

[19] Hall LL, Th'ng JP, Guo XW, Teplitz RL, Bradbury EM. A brief staurosporine treatment of mitotic cells triggers premature exit from mitosis and polyploid cell formation. Cancer Res. 1996; 56(15):3551–3559.

[20] Waldman T, Lengauer C, Kinzler KW, Vogelstein B. Uncoupling of S phase and mitosis induced by anticancer agents in cells lacking p21. Nature. 1996; 381(6584):713–716.

[21] Schneider DO, Whitmore GF. Comparative effects of neutrons and x-rays on mammalian cells. Radiat Res. 1963; 18:286–306.

[22] Bartek J, Lukas J. Chk1 and Chk2 kinases in checkpoint control and cancer. Cancer Cell. 2003; 3:421–429.

[23] Malumbres M, Barbacid M. Mammalian cyclin-dependent kinases. Trends Biochem Sci. 2005; 30:630–41.

[24] Liu Q, Guntuku S, Cui XS, Matsuoka S, Cortez D, Tamai K, Luo G, Carattini-Rivera S, DeMayo F, Bradley A, Donehower LA, Elledge SJ. Chk1 is an essential kinase that is regulated by Atr and required for the G(2)/M DNA damage checkpoint. Genes Dev. 2000; 14(12):1448–1459.

[25] Chen MS, Ryan CE, Piwnica-Worms H. Chk1 kinase negatively regulates mitotic function of Cdc25A phosphatase through 14-3-3 binding. Mol Cell Biol. 2003; 23(21): 7488–7497.

[26] Lam MH, Liu Q, Elledge SJ, Rosen JM. Chk1 is haploinsufficient for multiple functions critical to tumor suppression. Cancer Cell. 2004; 6(1):45–59.

[27] Chehab NH, Malikzay A, Appel M, Halazonetis TD. Chk2/hCds1 functions as a DNA damage checkpoint in G(1) by stabilizing p53. Genes Dev. 2000; 14(3):278–288.

[28] Shieh SY, Ahn J, Tamai K, Taya Y, Prives C. The human homologs of checkpoint kinases Chk1 and Cds1 (Chk2) phosphorylate p53 at multiple DNA damage-inducible sites. Genes Dev. 2000; 14(3):289–300.

[29] Falck J, Mailand N, Syljuåsen RG, Bartek J, Lukas J. The ATM–Chk2–Cdc25A checkpoint pathway guards against radioresistant DNA synthesis. Nature. 2001; 410(6830): 842–847.

[30] Peng CY, Graves PR, Thoma RS, Wu Z, Shaw AS, Piwnica-Worms H. Mitotic and G2 checkpoint control: regulation of 14-3-3 protein binding by phosphorylation of Cdc25C on serine-216. Science. 1997; 277(5331):1501–1505.

[31] Roninson IB, Broude EV, Chang BD. If not apoptosis, then what? Treatment-induced senescence and mitotic catastrophe in tumor cells. Drug Resist Updat. 2001; 4(5):303–313.

[32] Lock RB, Stribinskiene L. Dual modes of death induced by etoposide in human epithelial tumor cells allow Bcl-2 to inhibit apoptosis without affecting clonogenic survival. Cancer Res. 1996 Sep 1; 56(17):4006–4012.

[33] Nabha SM, Mohammad RM, Dandashi MH, Coupaye-Gerard B, Aboukameel A, Pettit GR, Al-Katib AM. Combretastatin-A4 prodrug induces mitotic catastrophe in chronic lymphocytic leukemia cell line independent of caspase activation and poly(ADP-ribose) polymerase cleavage. Clin Cancer Res. 2002; 8(8):2735–2741.

[34] Susin SA, Lorenzo HK, Zamzami N, Marzo I, Snow BE, Brothers GM, Mangion J, Jacotot E, Costantini P, Loeffler M, Larochette N, Goodlett DR, Aebersold R, Siderovski DP, Penninger JM, Kroemer G. Molecular characterization of mitochondrial apoptosis-inducing factor. Nature. 1999; 397(6718):441–446.

[35] Susin SA, Daugas E, Ravagnan L, Samejima K, Zamzami N, Loeffler M, Costantini P, Ferri KF, Irinopoulou T, Prévost MC, Brothers G, Mak TW, Penninger J, Earnshaw WC, Kroemer G. Two distinct pathways leading to nuclear apoptosis. J Exp Med. 2000; 192(4):571–580.

[36] Joza N, Susin SA, Daugas E, Stanford WL, Cho SK, Li CY, Sasaki T, Elia AJ, Cheng HY, Ravagnan L, Ferri KF, Zamzami N, Wakeham A, Hakem R, Yoshida H, Kong YY, Mak TW, Zúñiga-Pflücker JC, Kroemer G, Penninger JM. Essential role of the mi-

tochondrial apoptosis-inducing factor in programmed cell death. Nature. 2001; 410(6828):549–554.

[37] Bunz F, Dutriaux A, Lengauer C, Waldman T, Zhou S, Brown JP, Sedivy JM, Kinzler KW, Vogelstein B. Requirement for p53 and p21 to sustain G2 arrest after DNA damage. Science. 1998; 282(5393):1497–1501.

[38] Chan TA, Hermeking H, Lengauer C, Kinzler KW, Vogelstein B. 14-3-3Sigma is required to prevent mitotic catastrophe after DNA damage. Nature. 1999; 401(6753): 616–620.

[39] Chen Z, Xiao Z, Chen J, Ng SC, Sowin T, Sham H, Rosenberg S, Fesik S, Zhang H. Human Chk1 expression is dispensable for somatic cell death and critical for sustaining G2 DNA damage checkpoint. Mol Cancer Ther. 2003; 2(6):543–548.

[40] Levesque AA, Eastman A. p53-based cancer therapies: Is defective p53 the Achilles heel of the tumor? Carcinogenesis. 2007; 28(1):13–20.

[41] Erenpreisa J, Kalejs M, Ianzini F, Kosmacek EA, Mackey MA, Emzinsh D, Cragg MS, Ivanov A, Illidge TM. Segregation of genomes in polyploid tumour cells following mitotic catastrophe. Cell Biol Int. 2005; 29(12):1005–1011.

[42] Chang BD, Broude EV, Fang J, Kalinichenko TV, Abdryashitov R, Poole JC, Roninson IB. p21Waf1/Cip1/Sdi1-induced growth arrest is associated with depletion of mitosis-control proteins and leads to abnormal mitosis and endoreduplication in recovering cells. Oncogene. 2000; 19(17):2165–2170.

[43] Eom YW, Kim MA, Park SS, Goo MJ, Kwon HJ, Sohn S, Kim WH, Yoon G, Choi KS. Two distinct modes of cell death induced by doxorubicin: apoptosis and cell death through mitotic catastrophe accompanied by senescence-like phenotype. Oncogene. 2005; 24(30):4765–4777.

[44] Mansilla S, Priebe W, Portugal J. Mitotic catastrophe results in cell death by caspase-dependent and caspase-independent mechanisms. Cell Cycle. 2006; 5(1):53–60.

[45] Vogel C, Hager C, Bastians H. Mechanisms of mitotic cell death induced by chemotherapy-mediated G2 checkpoint abrogation. Cancer Res. 2007; 67(1):339–345.

[46] Niida H, Tsuge S, Katsuno Y, Konishi A, Takeda N, Nakanishi M. Depletion of Chk1 leads to premature activation of Cdc2-cyclin B and mitotic catastrophe. J Biol Chem. 2005; 280(47):39246–39252.

[47] Stark GR, Taylor WR. Control of the G2/M transition. Mol Biotechnol. 2006 Mar; 32(3):227–248.

[48] Enders GH. Expanded roles for Chk1 in genome maintenance. J Biol Chem. 2008; 283(26):17749–17752.

[49] Pines J, Hunter T. Isolation of a human cyclin cDNA: evidence for cyclin mRNA and protein regulation in the cell cycle and for interaction with p34cdc2. Cell. 1989; 58(5): 833–846.

[50] Nigg EA. Mitotic kinases as regulators of cell division and its checkpoints. Nat Rev Mol Cell Biol. 2001; 2(1):21–32.

[51] Weaver BA, Cleveland DW. Decoding the links between mitosis, cancer, and chemotherapy: The mitotic checkpoint, adaptation, and cell death. Cancer Cell. 2005; 8(1):7–12.

[52] Wang X, Cheung HW, Chun AC, Jin DY, Wong YC. Mitotic checkpoint defects in human cancers and their implications to chemotherapy. Front Biosci. 2008; 13:2103–2114.

[53] Smits VA, Medema RH. Checking out the G(2)/M transition. Biochim Biophys Acta. 2001; 1519(1–2):1–12.

[54] Yu H. Regulation of APC–Cdc20 by the spindle checkpoint. Curr Opin Cell Biol. 2002; 14(6):706–714.

[55] Gallant P, Nigg EA. Cyclin B2 undergoes cell cycle-dependent nuclear translocation and, when expressed as a non-destructible mutant, causes mitotic arrest in HeLa cells. J Cell Biol. 1992; 117(1):213–224.

[56] Rieder CL, Maiato H. Stuck in division or passing through: what happens when cells cannot satisfy the spindle assembly checkpoint. Dev Cell. 2004; 7(5):637–651.

[57] Weaver BA, Cleveland DW. Aneuploidy: instigator and inhibitor of tumorigenesis. Cancer Res. 2007; 67(21):10103–10105.

[58] Sandell LL, Zakian VA. Loss of a yeast telomere: arrest, recovery, and chromosome loss. Cell. 1993; 75(4):729–739.

[59] Toczyski DP, Galgoczy DJ, Hartwell LH. CDC5 and CKII control adaptation to the yeast DNA damage checkpoint. Cell. 1997; 90(6):1097–1106.

[60] Bartek J, Lukas J. DNA damage checkpoints: from initiation to recovery or adaptation. Curr Opin Cell Biol. 2007; 19(2):238–245.

[61] Tao W, South VJ, Zhang Y, Davide JP, Farrell L, Kohl NE, Sepp-Lorenzino L, Lobell RB. Induction of apoptosis by an inhibitor of the mitotic kinesin KSP requires both activation of the spindle assembly checkpoint and mitotic slippage. Cancer Cell. 2005; 8(1):49–59.

[62] Brito DA, Rieder CL. Mitotic checkpoint slippage in humans occurs via cyclin B destruction in the presence of an active checkpoint. Curr Biol. 2006; 16(12):1194–1200.

[63] Kotani S, Tanaka H, Yasuda H, Todokoro K. Regulation of APC activity by phosphorylation and regulatory factors. J Cell Biol. 1999; 146(4):791–800.

[64] Tolmach LJ, Marcus PI. Development of X-ray induced giant HeLa cells. Exp Cell Res. 1960; 20:350–360.

[65] Castedo M, Perfettini JL, Roumier T, Valent A, Raslova H, Yakushijin K, Horne D, Feunteun J, Lenoir G, Medema R, Vainchenker W, Kroemer G. Mitotic catastrophe constitutes a special case of apoptosis whose suppression entails aneuploidy. Oncogene. 2004; 23:4362–4370.

[66] Skwarska A, Augustin E, Konopa J. Sequential induction of mitotic catastrophe followed by apoptosis in human leukemia MOLT4 cells by imidazoacridinone C-1311. Apoptosis. 2007 Dec; 12:2245–2257.

[67] DeLuca JG, Moree B, Hickey JM, Kilmartin JV, Salmon ED. hNuf2 inhibition blocks stable kinetochore–microtubule attachment and induces mitotic cell death in HeLa cells. J Cell Biol. 2002; 159:549–555.

[68] Jordan MA, Wendell K, Gardiner S, Derry WB, Copp H, Wilson L. Mitotic block induced in HeLa cells by low concentrations of paclitaxel (Taxol) results in abnormal mitotic exit and apoptotic cell death. Cancer Res. 1996; 56:816–825.

[69] Woods CM, Zhu J, McQueney PA, Bollag D, Lazarides E. Taxol-induced mitotic block triggers rapid onset of a p53-independent apoptotic pathway. Mol Med. 1995; 1:506–526.

[70] Lanni JS, Jacks T. Characterization of the p53-dependent postmitotic checkpoint following spindle disruption. Mol Cell Biol. 1998; 18:1055–1064.

[71] Lanni JS, Jacks T. Characterization of the p53-dependent postmitotic checkpoint following spindle disruption. Mol Cell Biol. 1998; 18:1055–1064.

[72] Ramel S, Sanchez CA, Schimke MK, Neshat K, Cross SM, Raskind WH, Reid BJ. Inactivation of p53 and the development of tetraploidy in the elastase-SV40 T antigen transgenic mouse pancreas. Pancreas. 1995; 11:213–222.

[73] Dewey WC, Ling CC, Meyn RE. Radiation-induced apoptosis: relevance to radiotherapy. Int J Radiat Oncol Biol Phys. 1995; 33:781–796.

[74] Feddersen RM, Ehlenfeldt R, Yunis WS, Clark HB, Orr HT. Disrupted cerebellar cortical development and progressive degeneration of Purkinje cells in SV40 T antigen transgenic mice. Neuron. 1992; 9:955–966.

[75] al-Ubaidi MR1, Hollyfield JG, Overbeek PA, Baehr W. Photoreceptor degeneration induced by the expression of simian virus 40 large tumor antigen in the retina of transgenic mice. Proc Natl Acad Sci USA. 1992; 89:1194–1198

[76] Lee EY Chang CY, Hu N, Wang YC, Lai CC, Herrup K, Lee WH, Bradley A. Mice deficient for Rb are nonviable and show defects in neurogenesis and haematopoiesis. Nature. 1992; 359:288–294.

[77] Freeman RS, Estus S, Johnson EM Jr. Analysis of cell cycle-related gene expression in postmitotic neurons: selective induction of cyclin D1 during programmed cell death. Neuron. 1994; 12(2):343–55.

[78] Vincent I, Rosado M, Davies P. Mitotic mechanisms in Alzheimer's disease? J Cell Biol. 1996; 132:413–425.

[79] Nagy Z, Esiri MM, Cato AM, Smith AD. Cell cycle markers in the hippocampus in Alzheimer's disease. Acta Neuropathol. 1997; 94:6–15

[80] Smith MZ, Nagy Z, Esiri MM. Cell cycle-related protein expression in vascular dementia and Alzheimer's disease. Neurosci Lett. 1999; 271:45–48.

[81] Busser J, Geldmacher DS, Herrup K. Ectopic cell cycle proteins predict the sites of neuronal cell death in Alzheimer's disease brain. J Neurosci. 1998; 18:2801–2807.

[82] Esteras N Bartolomé F, Alquézar C, Antequera D, Muñoz Ú, Carro E, Martín-Requero Á. Altered cell cycle-related gene expression in brain and lymphocytes from a transgenic mouse model of Alzheimer's disease [amyloid precursor protein/presenilin 1 (PS1)]. Eur J Neurosci. 2012; 36:2609–2618

[83] Bonda DJ, Evans TA, Santocanale C, Llosá JC, Viña J, Bajic VP, Castellani RJ, Siedlak SL, Perry G, Smith MA, Lee HG. Evidence for the progression through S-phase in the ectopic cell cycle re-entry of neurons in Alzheimer disease. Aging (Albany NY). 2009; 1:382–388.

[84] Raina AK, Zhu X, Rottkamp CA, Monteiro M, Takeda A, Smith MA. Cyclin' toward dementia: cell cycle abnormalities and abortive oncogenesis in Alzheimer disease. J Neurosci Res. 2000; 61:128–133.

[85] Ogawa O, Zhu X, Lee HG, Raina A, Obrenovich ME, Bowser R, Ghanbari HA, Castellani RJ, Perry G, Smith MA. Ectopic localization of phosphorylated histone H3 in Alzheimer's disease: a mitotic catastrophe? Acta Neuropathol. 2003; 105:524–528.

[86] Arendt T, Holzer M, Fruth R, Brückner MK, Gärtner U. Paired helical filament-like phosphorylation of tau, deposition of beta/A4-amyloid and memory impairment in rat induced by chronic inhibition of phosphatase 1 and 2A. Neuroscience. 1995; 69:691–698.

[87] Lee J, Hong H, Im J, Byun H, Kim D. The formation of PHF-1 and SMI-31 positive dystrophic neurites in rat hippocampus following acute injection of okadaic acid. NEUROSCI Lett. 2000; 282:49–52.

[88] Nuydens RJ, de Jong M, Van Den Kieboom G, Heers C, Dispersyn G, Cornelissen F, Nuyens R, Borgers M, Geerts H. Okadaic acid-induced apoptosis in neuronal cells: evidence for an abortive mitotic attempt. Neurochem. 1998; 70:1124–1133.

[89] Chen B, Cheng M, Hong DJ, Sun FY, Zhu CQ. Okadaic acid induced cyclin B1 expression and mitotic catastrophe in rat cortex. Neurosci Lett. 2006; 406:178–182.

[90] Zhu X, Siedlak SL, Wang Y, Perry G, Castellani RJ, Cohen ML, Smith MA. Neuronal binucleation in Alzheimer disease hippocampus. Neuropathol Appl Neurobiol. 2008; 34:457–465.

[91] Vincent I, Jicha G, Rosado M, Dickson D. Aberrant expression of mitotic Cdc2/cyclin B1 kinase in degenerating neurons of Alzheimer's disease brain. J Neurosci. 1997; 17:3588–3598.

[92] Pei JJ, Braak H, Gong CX, Grundke-Iqbal I, Iqbal K, Winblad B, et al. Up-regulation of cell division cycle (cdc) 2 kinase in neurons with early stage Alzheimer's disease neurofibrillary degeneration. Acta Neuropathol. 2002; 104:369–376.

[93] Currais A, Hortobágyi T, Soriano S. The neuronal cell cycle as a mechanism of pathogenesis in Alzheimer's disease. Aging (Albany NY) 2009; 1:363–371.

[94] Konishi Y, Lehtinen M, Donovan N, Bonni A. Cdc2 phosphorylation of BAD links the cell cycle to the cell death machinery. Mol Cell. 2002; 9:1005–1016.

[95] Ranganathan S, Bowser R. Alterations in G(1) to S phase cell-cycle regulators during amyotrophic lateral sclerosis. Am J Pathol. 2003; 162:823–835.

[96] Ranganathan S, Bowser R. p53 and cell cycle proteins participate in spinal motor neuron cell death in ALS. Open Pathol J. 2010; 4:11–22.

[97] Ferraiuolo L, Heath PR, Holden H, Kasher P, Kirby J, Shaw PJ. Microarray analysis of the cellular pathways involved in the adaptation to and progression of motor neuron injury in the SOD1 G93A mouse model of familial ALS. J Neurosci. 2007; 27:9201–9219.

[98] Yang YJ, Herrup K. Loss of neuronal cell cycle control in ataxia-telangiectasia: a unified disease mechanism. Neuroscience. 2005; 25:2522–2529.

[99] Jordan-Sciutto KL, Dorsey R, Chalovich EM, Hammond RR, Achim CL. Expression patterns of retinoblastoma protein in Parkinson disease. J Neuropathol Exp Neurol. 2003; 62:68–74.

[100] Höglinger GU, Breunig JJ, Depboylu C, Rouaux C, Michel PP, Alvarez-Fischer D, Boutillier AL, Degregori J, Oertel WH, Rakic P, Hirsch EC, Hunot S.. The pRb/E2F cell-cycle pathway mediates cell death in Parkinson's disease. Proc Natl Acad Sci USA. 2007; 104:3585–3590.

[101] El-Khodor BF, Oo TF, Kholodilov N, Burke RE. Ectopic expression of cell cycle markers in models of induced programmed cell death in dopamine neurons of the rat substantia nigra pars compacta. Exp Neurol. 2003; 179.17–27.

[102] Rodriguez-Blanco J, Martín V, Herrera F, García-Santos G, Antolín I, Rodriguez C. Intracellular signaling pathways involved in post-mitotic dopaminergic PC12 cell death induced by 6-hydroxydopamine. J Neurochem. 2008; 107:127–140.

[103] Lee HG, Casadesus G, Nunomura A, Zhu X, Castellani RJ, Richardson SL, Perry G, Felsher DW, Petersen RB, Smith MA. The neuronal expression of MYC causes a neu-

rodegenerative phenotype in a novel transgenic mouse. Am J Pathol. 2009; 174:891–897.

[104] Timsit S, Rivera S, Ouaghi P, Guischard F, Tremblay E, Ben-Ari Y, Khrestchatisky M. Increased cyclin D1 in vulnerable neurons in the hippocampus after ischaemia and epilepsy: a modulator of in vivo programmed cell death?. Eur J Neurosci. 1999; 11:263–278.

[105] Hayashi T, Sakai K, Sasaki C, Zhang WR, Abe K. Phosphorylation of retinoblastoma protein in rat brain after transient middle cerebral artery occlusion. Neuropathol Appl Neurobiol. 2000; 26:390–397

[106] Wen Y, Yang S, Liu R, Simpkins JW. Cell-cycle regulators are involved in transient cerebral ischemia induced neuronal apoptosis in female rats. FEBS Lett. 2005; 579:4591–4599.

[107] Rashidian J, Iyirhiaro G, Aleyasin H, Rios M, Vincent I, Callaghan S, Bland RJ, Slack RS, During MJ, Park DS. Multiple cyclin-dependent kinases signals are critical mediators of ischemia/hypoxic neuronal death *in vitro* and *in vivo*. Proc Natl Acad Sci USA. 2005; 102:14080–14085.

[108] Bossenmeyer-Pourie C, Lièvre V, Grojean S, Koziel V, Pillot T, Daval JL. Sequential expression patterns of apoptosis- and cell cycle-related proteins in neuronal response to severe or mild transient hypoxia. Neuroscience. 2002; 114:869–882.

[109] Tomasevic G, Kamme F, Wieloch T. Changes in proliferating cell nuclear antigen, a protein involved in DNA repair, in vulnerable hippocampal neurons following global cerebral ischemia. Brain Res Mol Brain Res. 1998 Oct 1; 60(2):168–176.

[110] Imai H, Harland J, McCulloch J, Graham DI, Brown SM, Macrae IM. Specific expression of the cell cycle regulation proteins, GADD34 and PCNA, in the peri-infarct zone after focal cerebral ischaemia in the rat. Eur J Neurosci. 2002 Jun; 15(12):1929–1936.

[111] Kato H, Takahashi A, Itoyama Y. Cell cycle protein expression in proliferating microglia and astrocytes following transient global cerebral ischemia in the rat. Brain Res Bull. 2003 May 15; 60(3):215–221.

[112] Nagy Z, Esiri MM. Neuronal cyclin expression in the hippocampus in temporal lobe epilepsy. Exp Neurol. 1988; 150:240–247.

[113] Park DS, Morris EJ, Padmanabhan J, Shelanski ML, Geller HM, Greene LA. Cyclin-dependent kinases participate in death of neurons evoked by DNA-damaging agents. J Cell Biol. 1998; 143:457–467.

[114] Park DS Levine B, Ferrari G, Greene LA. Cyclin dependent kinase inhibitors and dominant negative cyclin dependent kinase 4 and 6 promote survival of NGF-deprived sympathetic neurons. J Neurosci. 1997; 17:8975–8983.

[115] Shankland SJ, Pippin JW, Couser WG. Complement (C5b-9) induces glomerular epithelial cell DNA synthesis but not proliferation *in vitro*. Kidney Int. 1999; 56(2):538–548.

[116] Couser WG, Pippin JW, Shankland SJ. Complement (C5b-9) induces DNA synthesis in rat mesangial cells *in vitro*. Kidney Int. 2001; 59:905–912.

[117] Pippin JW, Durvasula R, Petermann A, Hiromura K, Couser WG, Shankland SJ. DNA damage is a novel response to sublytic complement C5b-9-induced injury in podocytes. J Clin Invest. 2003 Mar; 111(6):877–85. (Shankland SJ Kidney Int. 2006; 69:2131–2147.)

[118] Terada Y, Inoshita S, Nakashima O, Kuwahara M, Sasaki S, Marumo F. Cyclins and the cyclin-kinase system--their potential roles in nephrology. Nephrol Dial Transplant. 1998; 13:1913–1916

[119] Shankland SJ. Cell cycle regulatory proteins in glomerular disease. Kidney Int. 1999; 56:1208–1215.

[120] Shankland SJ. Cell-cycle control and renal disease. Kidney Int. 1997; 52:294–308.

[121] Barisoni L, Mokrzycki M, Sablay L, Nagata M, Yamase H, Mundel P. Podocyte cell cycle regulation and proliferation in collapsing glomerulopathies. Kidney Int. 2000; 58:137–143.

[122] Srivastava T, Garola RE, Singh HK. Cell-cycle regulatory proteins in the podocyte in collapsing glomerulopathy in children. Kidney Int. 2006; 70:529–535.

[123] Wang S, Kim JH, Moon KC, Hong HK, Lee HS. Cell-cycle mechanisms involved in podocyte proliferation in cellular lesion of focal segmental glomerulosclerosis. Am J Kidney Dis. 2004; 43:19–27.

[124] Shankland SJ, Eitner F, Hudkins KL, Goodpaster T, D'Agati V, Alpers CE. Differential expression of cyclin-dependent kinase inhibitors in human glomerular disease: role in podocyte proliferation and maturation. Kidney Int. 2000; 58:674–683.

[125] Hiromura K, Haseley LA, Zhang P, Monkawa T, Durvasula R, Petermann AT, Alpers CE, Mundel P, Shankland SJ. Podocyte expression of the CDK-inhibitor p57 during development and disease. Kidney Int. 2001; 60:2235–2246.

[126] Shankland SJ, Floege J, Thomas SE, Nangaku M, Hugo C, Pippin J, Henne K, Hockenberry DM, Johnson RJ, Couser WG. Cyclin kinase inhibitors are increased during experimental membranous nephropathy: potential role in limiting glomerular epithelial cell proliferation *in vivo*. Kidney Int. 1997; 52:404–413

[127] Hoshi S, Shu Y, Yoshida F, Inagaki T, Sonoda J, Watanabe T, Nomoto K, Nagata M. Podocyte injury promotes progressive nephropathy in zucker diabetic fatty rats. Lab Invest. 2002; 82:25–35.

[128] Nitta K, Horita S, Honda K, Uchida K, Watanabe T, Nihei H, Nagata M. Glomerular expression of cell-cycle-regulatory proteins in human crescentic glomerulonephritis. Virchows Arch. 1999; 435:422–427.

[129] Petermann AT, Pippin J, Durvasula R, Pichler R, Hiromura K, Monkawa T, Couser WG, Shankland SJ. Mechanical stretch induces podocyte hypertrophy *in vitro*. Kidney Int. 2005; 67:157–166.

[130] Lasagni L, Ballerini L, Angelotti ML, Parente E, Sagrinati C, Mazzinghi B, Peired A, Ronconi E, Becherucci F, Bani D, Gacci M, Carini M, Lazzeri E, Romagnani P. Notch activation differentially regulates renal progenitors proliferation and differentiation toward the podocyte lineage in glomerular disorders. Stem cells. 2010; 28:1674–1685.

[131] Hara M, Yanagihara T, Kihara I. Urinary podocytes in primary focal segmental glomerulosclerosis. Nephron. 2001; 89:342–347.

[132] Vogelmann SU, Nelson WJ, Myers BD, Lemley KV. Urinary excretion of viable podocytes in health and renal disease. Am J Physiol Renal Physiol. 2003; 285:F40–F48.

[133] Buttitta LA, Edgar BA. Mechanisms controlling cell cycle exit upon terminal differentiation. Curr Opin Cell Biol. 2007; 19:697–704.

[134] Nagata M, Yamaguchi Y, Komatsu Y, Ito K. Mitosis and the presence of binucleate cells among glomerular podocytes in diseased human kidneys. Nephron. 1995; 70:68–71.

[135] Nagata M, Nakayama K, Terada Y, Hoshi S, Watanabe T. Cell cycle regulation and differentiation in the human podocyte lineage. Am J Pathol. 1998; 153:1511–1520.

[136] Mulay SR, Thomasova D, Ryu M, Kulkarni OP, Migliorini A, Bruns H, Gröbmayr R, Lazzeri E, Lasagni L, Liapis H, Romagnani P, Anders HJ. Podocyte loss involves MDM2-driven mitotic catastrophe. J Pathol. 2013; 230:322–335.

[137] Liapis H, Romagnani P, Anders HJ. New insights into the pathology of podocyte loss: mitotic catastrophe. Am J Pathol. 2013; 183:1364–1374.

[138] Wyatt CM, Meliambro K, Klotman PE. Recent progress in HIV-associated nephropathy. Annu Rev Med. 2012; 63:147–159.

[139] Craigie R. The molecular biology of HIV integrase. Future Virol. 2012; 7:679–686.

[140] Barisoni L, Bruggeman LA, Mundel P, D'Agati VD, Klotman PE. HIV-1 induces renal epithelial dedifferentiation in a transgenic model of HIV-associated nephropathy. Kidney Int. 2000; 53:173–181.

[141] Niranjan T, Bielesz B, Gruenwald A, Ponda MP, Kopp JB, Thomas DB, Susztak K. The Notch pathway in podocytes plays a role in the development of glomerular disease. Nat Med. 2008; 14:290–298.

[142] Murea M, Park JK, Sharma S, Kato H, Gruenwald A, Niranjan T, Si H, Thomas DB, Pullman JM, Melamed ML, Susztak K. Expression of Notch pathway proteins corre-

lates with albuminuria, glomerulosclerosis, and renal function. Kidney Int. 2010; 78:514–522.

[143] Humpreys BD, Czerniak S, DiRocco DP, Hasnain W, Cheema R, Bonventre JV. Repair of injured proximal tubule does not involve specialized progenitors. Proc Natl Acad Sci USA. 2011, 108:9226–9231.

[144] Humpreys BD, Valerius MT, Kobayashi A, Mugford JW, Soeung S, Duffield JS, McMahon AP, Bonventre JV. Intrinsic epithelial cells repair the kidney after injury. Cell Stem Cell. 2008; 2:284–291.

[145] Megyesi J, Safirstein RL, Price PM. Induction of p21WAF1/CIP1/SDI1 in kidney tubule cells affects the course of cisplatin-induced acute renal failure. J Clin Invest. 1997; 101:777–782.

[146] Megyesi J, Andrade L, Vieira JM Jr, Safirstein RL, Price PM. Positive effect of the induction of p21WAF1/CIP1 on the course of ischemic acute renal failure. Kidney Int. 2001; 60:2164–2172.

[147] Witzgall R, Brown D, Schwarz C, Bonventre JV. Localization of proliferating cell nuclear antigen, vimentin, c-Fos, and clusterin in the postischemic kidney. Evidence for a heterogenous genetic response among nephron segments, and a large pool of mitotically active and dedifferentiated cells. J Clin Invest. 1994; 93:2175–2188.

[148] Safirstein R, Price PM, Saggi SJ, Harris RC. Changes in gene expression after temporary renal ischemia. Kidney Int. 1990; 37:1515–1521.

[149] Ouellette AJ, Malt RA, Sukhatme VP, Bonventre JV. Expression of two "immediate early" genes, Egr-1 and c-fos, in response to renal ischemia and during compensatory renal hypertrophy in mice. J Clin Invest. 1990; 85:766–771.

[150] Price PM, Safirstein RL, Megyesi J. Protection of renal cells from cisplatin toxicity by cell cycle inhibitors. Am J Physiol Renal Physiol. 2004; 286:F378–F384.

[151] Megyesi J, Udvarhelyi N, Safirstein RL, Price PM. The p53-independent activation of transcription of p21 WAF1/CIP1/SDI1 after acute renal failure. Am J Physiol. 1995; 271:F1211–F1216.

[152] Price PM, Megyesi J, Saf Irstein RL. Cell cycle regulation: repair and regeneration in acute renal failure. Kidney Int. 2004; 66:509–514.

[153] Miyaji T, Kato A, Yasuda H, Fujigaki Y, Hishida A. Role of the increase in p21 in cisplatin-induced acute renal failure in rats. J Am Soc Nephrol. 2001; 12:900–908.

[154] Inguaggiato P, Gonzalez-Michaca L, Croatt AJ, Haggard JJ, Alam J, Nath KA. Cellular overexpression of heme oxygenase-1 up-regulates p21 and confers resistance to apoptosis. Kidney Int. 2001; 60:2181–2191.

[155] Golzalez-Michaca L, Farrugia G, Croatt AJ, Alam J, Nath KA. Heme: a determinant of life and death in renal tubular epithelial cells. Am J Physiol Renal Physiol. 2004 Feb; 286(2):F370–F377.

[156] Price PM, Yu F, Kaldis P, Aleem E, Nowak G, Safirstein RL, Megyesi J. Dependence of cisplatin-induced cell death *in vitro* and *in vivo* on cyclin-dependent kinase 2. J Am Soc Nephrol. 2006; 17:2434–2442.

[157] Hodeify R, Megyesi J, Tarcsafalvi A, Safirstein RL, Price PM. Protection of cisplatin cytotoxicity by an inactive cyclin-dependent kinase. Am J Physiol Renal Physiol. 2010 Jul; 299(1):F112–F120.

[158] Johnson SM, Torrice CD, Bell JF, Monahan KB, Jiang Q, Wang Y, Ramsey MR, Jin J, Wong KK, Su L, Zhou D, Sharpless NE. Mitigation of hematologic radiation toxicity in mice through pharmacological quiescence induced by CDK4/6 inhibition. J Clin Invest. 2010; 120:2528–2536.

[159] DiRocco DP, Bisi J, Roberts P, Strum J, Wong KK, Sharpless N, Humphreys BD. CDK4/6 inhibition induces epithelial cell cycle arrest and ameliorates acute kidney injury. Am J Physiol Renal Physiol. 2014; 306:F379–F388.

[160] Meersch M, Schmidt C, Van Aken H, Martens S, Rossaint J, Singbartl K, Görlich D, Kellum JA, Zarbock A. Urinary TIMP-2 and IGFBP7 as early biomarkers of acute kidney injury and renal recovery following cardiac surgery. PlosOne. 2014; 9:e93460.

[161] Kashani K, Al-Khafaji A, Ardiles T, Artigas A, Bagshaw SM, Bell M, Bihorac A, Birkhahn R, Cely CM, Chawla LS, Davison DL, Feldkamp T, Forni LG, Gong MN, Gunnerson KJ, Haase M, Hackett J, Honore PM, Hoste EA, Joannes-Boyau O, Joannidis M, Kim P, Koyner JL, Laskowitz DT, Lissauer ME, Marx G, McCullough PA, Mullaney S, Ostermann M, Rimmelé T, Shapiro NI, Shaw AD, Shi J, Sprague AM, Vincent JL, Vinsonneau C, Wagner L, Walker MG, Wilkerson RG, Zacharowski K, Kellum JA. Discovery and validation of cell cycle arrest biomarkers in human acute kidney injury. Crit Care. 2013; 17:R25.

[162] Yang L, Besschetnova TY, Brooks CR, Shah JV, Bonventre JV. Epithelial cell cycle arrest in G2/M mediates kidney fibrosis after injury. Nat Med. 2010; 16:535.

[163] Wu CF, Chiang WC, Lai CF, Chang FC, Chen YT, Chou YH, Wu TH, Linn GR, Ling H, Wu KD, Tsai TJ, Chen YM, Duffield JS, Lin SL. Transforming growth factor β-1 stimulates profibrotic epithelial signaling to activate pericyte–myofibroblast transition in obstructive kidney fibrosis. Am J Pathol. 2013; 182:118–131.

[164] Tang J, Liu N, Tolbert E, Ponnusamy M, Ma L, Gong R, Bayliss G, Yan H, Zhuang S. Sustained activation of EGFR triggers renal fibrogenesis after acute kidney injury. Am J Pathol. 2013; 183:160–172.

[165] Wu CF, Chiang WC, Lai CF, Chang FC, Chen YT, Chou YH, Wu TH, Linn GR, Ling H, Wu KD, Tsai TJ, Chen YM, Duffield JS, Lin SL. Transforming growth factor β-1

stimulates profibrotic epithelial signaling to activate pericyte–myofibroblast transition in obstructive kidney fibrosis. Am J Pathol. 2013; 182:118–131.

[166] Cianciolo C, Skrypnyk NI, Brilli LL, Chiba T, Novitskaya T, Woods C, West J, Korotchenko VN, McDermott L, Day BW, Davidson AJ, Harris RC, de Caestecker MP, Hukriede NA. Histone deacetylase inhibitor enhances recovery after AKI. JASN. 2013; 24:943–953.

[167] Zhou W, Otto EA, Cluckey A, Airik R, Hurd TW, Chaki M, Diaz K, Lach FP, Bennett GR, Gee HY, Ghosh AK, Natarajan S, Thongthip S, Veturi U, Allen SJ, Janssen S, Ramaswami G, Dixon J, Burkhalter F, Spoendlin M, Moch H, Mihatsch MJ, Verine J, Reade R, Soliman H, Godin M, Kiss D, Monga G, Mazzucco G, Amann K, Artunc F, Newland RC, Wiech T, Zschiedrich S, Huber TB, Friedl A, Slaats GG, Joles JA, Goldschmeding R, Washburn J, Giles RH, Levy S, Smogorzewska A, Hildebrandt F. FAN1 mutations cause karyomegalic interstitial nephritis, linking chronic kidney failure to defective DNA damage repair. Nature Genet. 2012; 44:910–915.

[168] Zhou L, Fu P, Huang XR, Liu F, Lai KN, Lan HY. Activation of p53 promotes renal injury in acute aristolochic acid nephropathy. JASN. 2010; 21:31–41.

Section 3

Cellular Basis of Disease and Therapy

Chapter 4

New Frontiers in Cancer Chemotherapy — Targeting Cell Death Pathways

Yong Zhong Xu, Cynthia Kanagaratham, Mina Youssef and Danuta Radzioch

Additional information is available at the end of the chapter

http://dx.doi.org/10.5772/61260

Abstract

Cell death plays an important role in tumorigenesis, growth, and progression and affects the efficiency of chemotherapy to a great extent. Apoptosis is usually regarded as the principal mechanism of chemotherapy-induced cell death. However, the dysregulation of apoptosis occurs commonly in many cancers, which lowers the effectiveness of therapy and allows cells to survive. The mechanisms by which cells acquire this resistance to chemotherapy are not fully understood. Several studies uncovered alternative cell death pathways that are mechanistically distinct from apoptosis. These pathways, including autophagy and necrosis, represent potential targets for novel cancer treatment. By modulating the key regulatory molecules involved in the different types of cell death, more effective and less toxic chemotherapy might be developed. In this chapter, we describe the signaling pathways and the molecular events that are involved in these three major forms of programmed cell death. Additionally, we also discuss the emerging therapies targeting these cell death pathways as new strategies against cancer.

Keywords: Cancer, apoptosis, autophagy, necrosis, targeted therapy

1. Introduction

According to the World Health Organization, cancers figure among the leading causes of death worldwide, accounting for 8.2 million deaths in 2012 [1]. The annual number of cancer cases are expected to rise from 14 million in 2012 to 22 million within the next two decades [1]. Cancer can affect everyone - the young and the old, the rich and the poor, men and women - and poses a tremendous burden on patients, families, and societies. A substantial number of cancer patients experience a significant reduction in their quality of life due to physical pain, mental anguish, and economic hardship. Scientists and doctors are continuously making efforts to find better and more effective therapies against cancer. Currently, strategically targeted cancer therapies are emerging as treatments, which use drugs or other substances, such as tyrosine kinase inhibitors and apoptosis inducing agents, to interfere with specific molecules and processes involved in cancer cell growth and survival [2].

In multicellular organisms, the number of cells is tightly regulated to attain a balance between cell proliferation and death. Maintaining this balance is crucial for normal development and tissue size homeostasis [3]. Cell death is a fundamental process that not only plays a pivotal role in the regulation of normal physiological development and tissue balance but also acts as a defense mechanism against diseases such as cancer [4]. Over the past two decades, our knowledge of cell death and the mechanisms of its regulation have increased dramatically. Programmed cell death (PCD) is a principal mechanism of tumor suppression and is triggered in nonmalignant cells to eliminate unnecessary, aged, or damaged cells that may otherwise be harmful to the body [5]. Of note, apoptosis, autophagy, and programmed necrosis are the three main forms of PCD, easily distinguished by their morphological characteristics within the cell [6, 7]. Additionally, senescence and mitotic catastrophe (MC) are two other cell death mechanisms, often triggered in cancer cells and tissues in response to anticancer drugs [8]. Cell senescence, a state of permanent cell-cycle arrest characterized by specific changes in morphology and gene expression that differentiate it from reversible cell cycle arrest, is also considered as a type of cell death in the context of cancer therapy [9].

Accumulated data suggest that various chemotherapeutic agents can kill tumor cells through the induction of apoptosis [10]. Dysregulation of the apoptotic pathways can not only promote tumorigenesis [11, 12] but also render cancer cells resistant to chemotherapy. The ability of cancer cells to avoid apoptosis and continue to proliferate is one of the fundamental hallmarks of cancer and is a major target of cancer therapy development [12].Development of novel molecules that activate apoptosis by targeting both the intrinsic and extrinsic apoptotic pathways will advance our understanding of the mechanisms behind tumor cell proliferation, which may also lead to the development of effective cancer therapies. Autophagy is an evolutionarily conserved process that maintains cellular homeostasis by controlling protein and organelle turnover. It serves as critical adaptive response that recycles energy and nutrients during periods of starvation and stress to enable cell survival. Studies have shown that autophagy contributes to the adaptation of tumor cells to adverse microenvironments [13] and chemotherapy [13]. Autophagy may represent a major impediment to successful cancer therapy; therefore, targeting autophagy is considered a promising strategy in clinical cancer treatment. However, other studies have shown that deficiency in adequate autophagy results in various spontaneous tumors in mouse model [14], indicating a tumor suppressive role of autophagy in the process of tumorigenesis. It seems that autophagy plays dual roles as both promoter and suppressor in tumorigenesis. The dynamic role of autophagy in tumor development appears mainly dependent on tumor stage [15]. It is important to elucidate the mechanisms by which autophagy influences tumorigenesis and treatment response. Analysis of autophagic signaling may identify novel therapeutic targets. Necrosis is generally considered a passive response to massive cellular damage. However, accumulating evidence supports the existence of programmed necrosis, which involves cell swelling, organelle dysfunction, and cell lysis [16, 17]. Given the fact that many cancers have defective apoptosis machinery, it is reasonable to consider the pros and cons of activating other cell death pathways, such as necrosis, senescence and MC, and assess their therapeutic potential.

In this chapter, we discuss three major forms of PCD at molecular, cellular, and physiological levels. We also discuss the regulation mechanisms of these cell death pathways. Finally, the emerging therapies and strategies targeting these cell death pathways in the treatment of cancers are examined.

2. Apoptosis

The term "apoptosis" originates from Greek words *apo*, which means "since, " and *ptosis*, which means "dropping off, " and it refers to leaves falling off trees or petals dropping off flowers. It was first coined by Kerr *et al.* in 1972 and used to describe a regulated form of cell death with specific morphological features, which is different from the necrotic cell death resulting from acute tissue injury [18]. Since then, apoptosis has become one of the most extensively studied forms of PCD that plays a critical role in normal biological processes, such as embryonic development, immune response, tissue homeostasis, and cell turnover [19], as well as in a variety of pathological conditions including cancer [20]. Studies have shown that a cell undergoing apoptosis can be described by a series of characteristic morphological changes, including cell shrinkage, membrane blebbing, chromatin condensation, and nuclear fragmentation [21]. In addition to the morphological changes, biochemical changes happening during apoptosis have also been revealed, and the three main ones are (1) the activation of caspases, (2) the breakdown of DNA and protein, and (3) the modifications of cell surface markers tagging the apoptotic cells for recognition by phagocytic cells [22].

Tissue homeostasis is maintained by an elaborate balance between cell growth by proliferation and/or survival on one side and cell death via apoptosis and other pathways on the other side. Any changes in the contribution of cell growth versus cell death can seriously affect the tissue homeostasis leading to human diseases. Accumulated evidence indicates that defect in apoptosis can contribute to cancer or onset of autoimmune responses, while excessive cell death can cause acute or chronic degenerative diseases, immunodeficiency, and infertility [23]. Under normal conditions, apoptosis represents a safeguard mechanism to prevent tumorigenesis, which indicates that evasion or resistance to apoptosis is a pivotal feature of cancer [24]. Alterations in cancer cells, which lead to impaired apoptotic signaling, not only promote tumor formation, progression, and metastasis but also contribute to treatment resistance [6-8, 24]. Thus, a better understanding of the molecular events that are involved in the regulation of apoptosis and their dysregulation in human cancers is expected to provide novel strategies for cancer therapy.

2.1. Two main signaling pathways involved in drug-induced apoptosis

Apoptosis can be triggered by various stimuli from outside or inside of the cells, for example, by ligation of cell surface receptors, by DNA damage as a results of treatment with cytotoxic drugs or irradiation, by a lack of survival signals or by developmental death signals. These death signals of diverse origins eventually converge to activate a series of cysteine aspartyl-specific proteases (caspases) through two main pathways, namely, extrinsic (death receptor) and intrinsic (mitochondrial) pathways [25]. Caspases are central to the mechanisms of apoptosis, which cleave key cellular proteins and dismantle the cells (Figure 1). Given the death-causing effects of caspase activation, these two pathways are strictly and closely regulated at each step. Apart from the two pathways mentioned, endoplasmic reticulum (ER)-mediated apoptosis is a lesser known third pathway [26].

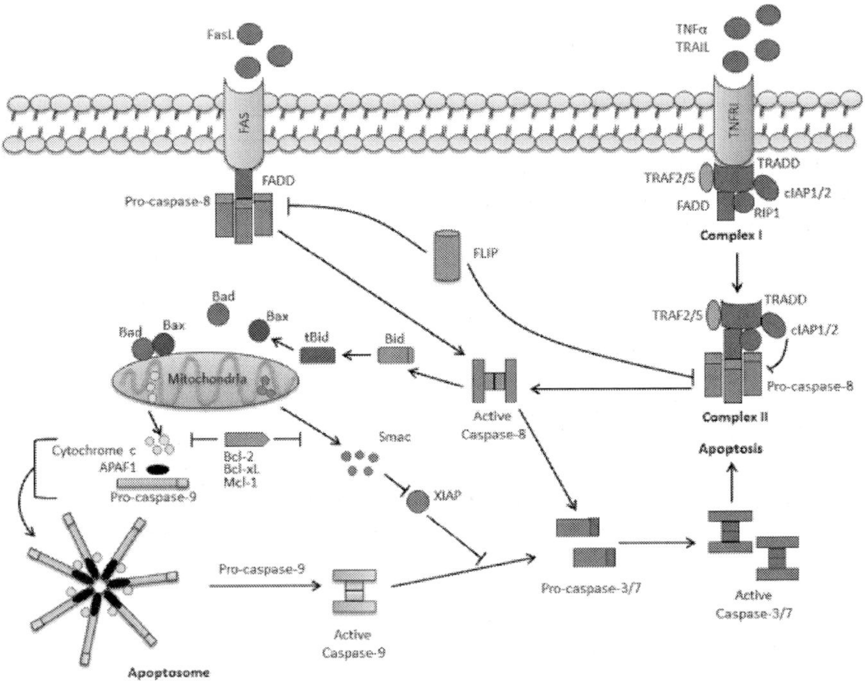

Figure 1. The extrinsic and intrinsic apoptosis signaling pathways. The extrinsic pathway primarily involves the activation of procaspase 8 by death receptors (e.g., TNFR1 and Fas/Apo 1), whereas the intrinsic pathway involves the release of factors from mitochondria, such as cytochrome *c*, that forms a complex with APAF1 and procaspase 9, resulting in the cleavage and activation of procaspase 9. In mammals, either active caspase 8 or caspase 9 is capable of activating effector caspases such as caspase 3 or caspase 7, which then cleave apoptotic substrates leading to apoptosis. A link between the extrinsic and intrinsic pathways is observed in certain cells. This involves the cleavage of the Bcl-2 family member Bid by caspase 8, leading to the release of cytochrome *c* from the mitochondria and activation of caspase 9. For detailed signaling pathways, please see Sections 2.1.1-2.1.3.

Apoptosis in response to cancer therapy proceeds through the activation of the core apoptotic machinery, including the receptor and the mitochondrial signaling pathway [10]. In many tumor cell types, the main signaling pathway leading from drug-induced damage to cell death involves the mitochondrial release of proapoptotic molecules under the control of the B-cell lymphoma 2 (Bcl-2) family of proteins. However, death receptors of the tumor necrosis factor receptor (TNFR) superfamily, mainly CD95 (APO-1/Fas), have also been shown to play a role in linking drug-induced damage to the apoptotic machinery and modulating drug response.

2.1.1. Extrinsic (death receptor) pathway

The extrinsic apoptotic pathway is activated by the binding of death ligands to cell surface death receptors, which transmit extracellular death signals to the intracellular apoptotic machinery to elicit cell death [27]. Although several death receptors have been identified, the best known death receptors belong to the TNFR superfamily, including TNFRs, CD95 (Fas/Apo 1), and tumor necrosis factor-related apoptosis-inducing ligand (TRAIL) receptors [28]. These receptors become activated once bound by their cognate ligands such as TNF, CD95 (Fas), and TRAIL, which in turn results in death receptor aggregation, and recruitment of various adaptor proteins to the intracellular death domains (DD) of the death receptors and formation of death-inducing signaling complex (DISC). In this complex, Fas-associated death domain (FADD) recruits other DD- and/or death effector domain (DED)-containing proteins, such as procaspase 8 and procaspase 10, via homotypic death domain interactions (Figure 1) [29]. In contrast, TNF receptor-associated death domain (TRADD) recruits proteins leading to the formation of two complexes [30]. For example, TNFR1 binds to and forms complex I with TRADD, TNF receptor-associated factors 2 and 5 (TRAF2/5), receptor interacting protein 1 (RIP1), and the cellular inhibitor of apoptosis proteins 1 and 2 (cIAP1/2) (Figure 1). This complex is important for the TNF-induced activation of NF-κB and MAPKs and the subsequent transcription of antiapoptotic genes. In certain circumstance, RIP1 is deubiquitinated by cylindromatosis and leads to the dissociation of RIP1 and TRADD from complex I. RIP1 and TRADD then form complex II with FADD, caspase 8, and/or caspase 10, which is analogous to the DISC induced by FasL and TRAIL (Figure 1) [4, 31, 32]. The activation of caspases 8 and 10 leads to the activation of the downstream caspase cascade to mediate apoptosis. In some cells, named type I cells, the activation of effector caspases, such as caspases 3, 6, and 7, by caspase 8/10 alone can induce apoptosis [33]. However, in type II cells, activated caspase 8/10 triggers the activation of intrinsic apoptotic pathway by the cleavage of the Bcl-2-homology 3 (BH3)-only protein Bid. Cleaved Bid is myristoylated to form tBid and translocates to the mitochondria membrane, which promotes the oligomerization of Bax and Bak and causes the release of apoptotic mediators from the mitochondria (Figure 1) [34, 35].

2.1.2. Regulation of the extrinsic pathway

Caspase 8 is the predominant initiator caspase in the extrinsic pathway, which plays a pivotal role in determining the cell fate following the death receptor activation. Therefore, the major signals that affect the recruitment of caspase 8 and its activation can modulate this signaling pathway. For example, cellular FADD-like interleukin-1β-converting enzyme inhibitory protein (cFLIP) shares significant structural similarities with caspases 8 and 10, which allows it to compete for binding sites and thus displace caspase 8/10 in the DISC complex. cFLIP lacks a functional caspase domain, suggesting it to be a dominant-negative inhibitor [36]. Besides caspase 8/10, cFLIP can also bind to FADD and TRAIL receptor 5 (DR5), and this interaction in turn prevents the formation of the DISC complex and the subsequent activation of caspase cascade [37]. Similarly, A20-binding inhibitor of NF-κB 1 (ABIN1) exerts its antiapoptotic effect by interfering with the interaction of RIP1 and FADD with caspase 8 [38].

cIAP1/2 contain a signature baculovirus IAP repeat (BIR), a caspase-recruitment domain (CARD), and a really interesting new gene (RING) domain at their C-terminal that exhibits E3 ubiquitin ligase activity, which help to recruit TRAF1/2 and inhibit TNFα-apoptotic signaling. Although cIAP1/2 are not efficient caspase 8 inhibitors, they can play a regulatory role in extrinsic pathway through the activation of prosurvival signals, such as NF-κB pathway. This effect was shown to result from the cIAP1/2 induction of RIP1 ubiquitination and the recruitment of TAK1, TAB2/3, and the IKK complex [39]. The NF-κB signaling pathway has been linked to death receptor signaling because RIP, which serves as an adaptor molecule for TNFR1 in the NF-κB pathway, can be cleaved by caspases. Upon TNF receptor signaling, this modulates the balance between proapoptotic and antiapoptotic signals and may even stimulate an autocrine "death loop" [10, 40]

Ubiquitination has been shown to regulate the activity of caspase 8. A clear example is that the polyubiquitination of the p10 subunit of caspase 8 by a cullin3-based E3 ligase can enhance its enzymatic activity [41]. This modification occurs after the recruitment of caspase 8 to DISC complex and allows for the binding of active caspase 8 to the polyubiquitination-binding protein, p62, which is thought to increase the stability of cleaved caspase 8 [41]. The deubiquitinating (DUB) enzyme A20 was reportedly involved in reversing this modification [41].

2.1.3. *Intrinsic (mitochondrial) pathway*

The intrinsic pathway, as implied by its name, is activated by internal stimuli such as DNA damaging agents, growth factor deprivation, oxidants, hypoxia, overload of calcium, and microtubule targeting drugs [42]. Upon the detection of the internal stimuli, two proapoptotic Bcl-2 family members, Bax and Bak, undergo structural changes and subsequent oligomerization at the outer membrane of the mitochondria, leading to the induction of mitochondrial outer membrane permeabilization (MOMP) and the release of mitochondrial cytochrome *c* (Cyt-c) into the cytosol [43-46]. The released Cyt-c assembles a multiprotein caspase-activating complex, known as the "apoptosome" [47]. The central component of the apoptosome is Apaf1 that is transiently bound by released Cyt-c in the presence of ATP or dATP, which leads to the oligomerization of Apaf1 and then the exposure of its CARD [48]. Subsequently, Apaf1 binds to procaspase 9 via interaction between their CARDs. In this complex, procaspase 9 dimerizes and autoactivates. Activated caspase 9 then cleaves and activates the downstream executioner caspases 3 and 7 to perpetrate cell death rapidly (Figure 1) [49]. Besides Cyt-c, other apoptotic factors are also released from the mitochondrial intermembrane space into the cytoplasm such as apoptosis-inducing factor (AIF), second mitochondria-derived activator of caspase/direct inhibitor of apoptosis (IAP)-binding protein with low pI (Smac/DIABLO), and Omi/high temperature requirement protein A2 (Omi/HtrA2) [50]. Smac/DIABLO and Omi/HtrA2 promote caspase activation by neutralizing the inhibitory effects on IAPs, while AIF causes DNA condensation [51-53]

2.1.4. Regulation of the intrinsic pathway

The intrinsic pathway is tightly regulated by the intricate interactions between pro- and antiapoptotic members of the Bcl-2 family, which are categorized according to the organization of their Bcl-2 homology (BH) domains: (1) antiapoptotic members such as Bcl-2, Bcl-xL, Bcl-w, A1, and Mcl-1, which all possess the four BH1-BH4 domains and inhibit proapoptotic counterparts; (2) effector proapoptotic members such as Bax, Bak, and Bok, which all possess the three domains BH1-BH3; and (3) BH3-only proteins, including Bid, Bad, Bim, Bik, Bmf, Hrk, Noxa, Puma, Blk, BNIP3, and Spike, which only have the short BH3 motif and promote MOMP, either by inhibiting antiapoptotic proteins or by activating Bax and Bak [54-56]. Antiapoptotic Bcl-2 members block the oligomerization of Bax and Bak or their association with BH3-only proteins, thus preventing MOMP and Cyt-c release. However, upon a cytotoxic stimulus, the effects of antiapoptotic members are counteracted by BH3-only proteins, such as Bim and Noxa. BH3-only proteins release Bax-Bak from inhibition and allow them to promote MOMP and apoptosis.

The inhibitor of apoptosis protein (IAP) family represents another negative regulator of the intrinsic apoptotic pathway. So far, eight members have been identified, including cIAP1, cIAP2, X-linked IAP (XIAP), neuronal apoptosis inhibitory protein (NAIP), melanoma IAP (ML-IAP), survivin, Apollon, and IAP like protein 2 (ILP2) [57]. All IAPs contain BIR domains and 70 amino acid motifs, which are essential for antiapoptotic properties of IAPs because BIR domains bind the active sites of caspases and inhibit proteolytic function. Indeed, XIAP, survivin, and cIAP1/2 have been found to directly inhibit caspases 3, 7, and 9 [58]. In case of XIAP, its BIR3 domain directly binds to the small subunit of caspase 9, while its BIR2 domain interacts with the active-site substrate-binding pocket of caspases 3 and 7 [59, 60]. Some IAPs such as cIAP1/2 and XIAP contain a highly conserved RING domain that possesses E3 ubiquitin ligase activity and may target effector caspases for ubiquitination and subsequent proteasomal degradation [61, 62].

Other apoptotic factors, for example, Smac/DIABLO, when released from the mitochondrial intermembrane space during mitochondrial apoptotic events, are able to bind to various IAPs, mainly XIAP in a manner that displaces caspases from XIAP and enables their activation. In addition, the binding of Smac/DIABLO to IAPs facilitates the latter to be degraded by proteasome [63]. However, unlike Cyt-c, the ablation of Smac/DIABLO, Omi/HtrA2, or both proteins does not lead to the inability to activate caspases or undergo apoptosis [64-66]. This suggests that there may be considerable redundancy in XIAP inhibition, and in fact other proteins have also been demonstrated to inhibit XIAP [67].

3. Autophagy

The term autophagy (from the Greek *auto*, meaning "oneself, " and *phagy*, meaning "eating") was first introduced by Christian de Duve based on the observation that cells were able to digest their own components [68]. Nowadays, autophagy is defined as a self-digestive cellular process by which eukaryotes degrade and recycle long-lived proteins, cellular aggregates, and damaged cellular organelles to maintain cellular homeostasis. Three types of autophagy, macroautophagy, microautophagy, and chaperone-mediated autophagy (CMA), have been identified to date, which could be distinguished from one another via different modes of delivery of the cargo to the lysosome and their function. Microautophagy is a direct engulfment of cytoplasmic components into the lysosomal lumen for degradation [69], while in CMA, a subset of soluble cytosolic proteins containing a KFERQ motif are recognized by molecular chaperons, including the HSPA8/HSC70 (heat shock 70 kDa protein 8), and directly translocated into lysosomes through a receptor (LAMP-2A) on the lysosomal membrane [70]. Macroautophagy is a process that is responsible for the delivery and degradation of macromolecules and organelles by generating specialized cytosolic vesicles (hereafter referred to as autophagy) [71].

Autophagy is activated under physiological and pathological conditions, such as nutrient starvation, hypoxia, hyperthermia, and oxidative stress, and in response to drugs and radiation. This dynamic process generates cellular energy resources that allow a cell to adapt its metabolism to energy demand. Defects in the autophagy process lead to the accumulation of damaged proteins and/or genomic damage and can cause diseases such as neurodegeneration, infectious diseases, heart diseases, and cancers [72, 73]. Although autophagy can suppress tumor growth, it clearly plays a role in promoting tumor cells to survive under stress [74]. The suppression of autophagy can sensitize cancer cells to anticancer therapy [75, 76], but under apoptosis deficiency condition, autophagy can also cause cell death through a process termed "autophagic cell death" [72, 77].

3.1. Pathway of autophagy

After induction by a stress signal such as starvation, the process of autophagy begins with the formation of autophagosomes, which assemble around and encapsulate the targeted proteins or organelles, and then fuse with lysosomes to form autolysosomes for degradation. This complex process can be divided into five major steps, namely, nucleation, elongation, maturation, fusion, and degradation, which are tightly controlled by a subset of molecules encoded by autophagy-related genes (ATGs) (Figure 2A). The first step of autophagy is the nucleation of the phagophore, an isolation membrane that most likely derives from the ER [78, 79]. Besides the ER, studies have also shown that the plasma membrane and membranes of mitochondria and Golgi are also involved in the formation of the phagophore [80, 81]. The phagophore then extends and sequesters the substrates destined for degradation and finally forms the characteristic double membrane vesicle, known as autophagosome. The outer membrane of the mature autophagosome then fuses with the lysosome or inner body to generate a structure named autolysosome, where the inner membrane of autophagosome and its contents are degraded by the activity of acidic hydrolases provided by the lysosome [82, 83]. The catabolic products are then either recycled into different metabolic pathways or undergo further degradation to yield energy (Figure 2A).

http://dx.doi.org/10.5772/61260

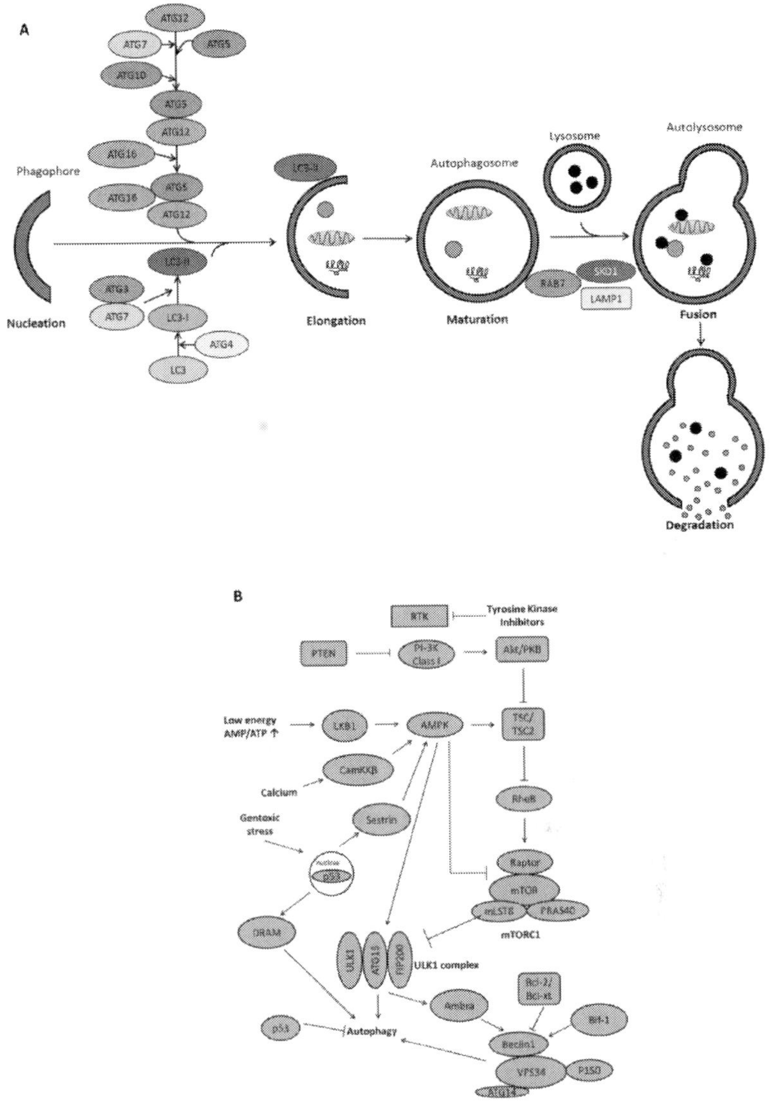

Figure 2. Schematic representations of the autophagy pathway and its regulation. (A) The five major steps of autophagy, namely, nucleation, elongation, maturation, fusion, and degradation, are illustrated. Phagophore membrane elongation and subsequent sealing of the autophagosome require two ubiquitin-like conjugation systems that mediate the formation of ATG5-ATG12 complex and LC3-II. (B) The signaling molecules and pathways involved in autophagy regulation (see Sections 3.1-3.3 for details).

Autophagy is a highly regulated process by ~30 ATGs discovered hitherto in mammals. Several signaling pathways that initiate autophagy converge at a serine/threonine protein kinase mammalian target of rapamycin (mTOR), a key regulator of the autophagic pathway, which inhibits autophagy in the presence of nutrients and growth factors [84]. In mammals, the initiation of phagophore formation is regulated by a great deal of macromolecular complexes or groups of proteins, including the ULK1 kinase and its regulators, the autophagy-specific phosphatidylinositol 3-kinase (PI3K) complex, and the multi-spanning transmembrane protein ATG9 [85-88]. The PI3K complex, which consists of the active enzyme VPS34, a class III PI3K, together with p150 and Beclin 1, the counterparts of yeast Vps15 and Vps30/Atg6, and ATG14, catalyze the production of phosphatidylinositol-3-phosphate, thereby triggering the recruitment of effectors proteins, such as double FYVE-containing protein 1 (DFCP1) and WD-repeat domain phosphoinositide-interacting (WIPI) family proteins [89-93]. The elongation of the isolation membrane and the subsequent closure of the autophagosome require two ubiquitin-like conjugation systems. First is the ATG12-ATG5-ATG16L system: ATG12 is conjugated to ATG5 by the ATG7 (E1-like enzyme) and ATG10 (E2-like enzyme). The resulting ATG5-ATG12 complex interacts with ATG16L and then oligomerizes to form a large ATG16L complex, which localizes on the outer surface of the extending autophagosomal membrane, but it dissociates from the membrane before autophagosome formation is completed (Figure 2A) [94]. A recent study demonstrated that under certain stress conditions, autophagy can occur independently of ATG5/ATG7, suggesting the existence of an alternative pathway for autophagosome formation [95]. Second is the phosphatidylethanolamine (PE)-light chain 3 (LC3) system: LC3 (the mammalian homologue of yeast Atg8) is cleaved by the cysteine protease ATG4 and then conjugated to the lipid PE by the activity of ATG7 and ATG3 (E2-like enzyme) [94, 96]. The lipidated form of LC3 (LC3II) specifically accumulates on nascent autophagosomes and recruits cargo adaptor proteins (also known as autophagy receptors), such as p62, Nbr1, or NIX. These proteins, in turn, recruit cargo from the cytoplasm, for example, ubiquitinated protein aggregates and damaged organelles, to promote the closure of the autophagosome [97-99]. Once autophagosome formation is complete, it fuses with lysosomes through mechanisms that remain largely unknown in mammalian cells. Some regulators have been found to be involved in the autophagosome-lysosome fusion process, including LC3, the lysosomal proteins LAMP-1 and LAMP-2, the small GTP-binding protein RAB7, and the AAA-type ATPase SKD1 [100-102]. Autophagosome-lysosome fusion then leads to the activation of the hydrolases and the degradation of the sequestered cargo (Figure 2A).

3.2. Signaling pathways involved in the regulation of autophagy

3.2.1. Mammalian target of rapamycin

The mTOR, a PI3K-related serine/threonine protein kinase, plays a key role in maintaining the balance between cell growth and proliferation. It has also been found to regulate the autophagy in response to nutrient status, growth factor signals, and cell stress [103]. In higher eukaryotes, mTOR exists in at least two distinct protein complexes, known as mTOR complex 1 (mTORC1) and mTOR complex 2 (mTORC2) [104, 105]. The mTORC1 consists of mTOR, mLST8 (mammalian lethal with SEC13 protein 8), RAPTOR (regulatory-associated protein of mTOR), PRAS40 (proline-rich Akt substrate of 40 kDa), and DEPTOR (DEP domain containing mTOR-interacting protein) and is considered to be the principal regulator of autophagy [106]. mTORC1 is downstream of PI3K and is activated in response to mitogenic stimuli or nutrient availability. When nutrients and growth factors are available, mTORC1 inhibits autophagy by phosphorylating ULK1-ATG13-FIP200 complex, which is required for the phagophore formation [107, 108]. In addition, activated mTORC1 promotes mRNA translation via activating S6K and inhibiting 4EBP1. The inhibition of mTORC1 strongly induces autophagy, for example, rapamycin (an mTORC1 inhibitor), and potently induces autophagy even in the presence of rich nutrients. As indicated (see Figure 2B), mTOR activity is activated by different signaling pathways, which converge on the tuberous sclerosis complex (TSC) and the ras homolog enriched in brain (Rheb), a small GTPase that activates mTORC1 when in its GTP-bound state [109]. TSC, which is comprised of TSC1 (harmartin) and TSC2 (tuberin), acts as a GTPase-activating protein (GAP) for Rheb, promoting hydrolysis of its bound GTP and thus inhibiting Rheb and mTORC1 activity [110]. The inhibition of TSC1/2 by Akt phosphorylation allows Rheb-GTP to accumulate and activates mTOR [111]. Phosphatase and tensin homology (PTEN), a phosphoinositide-3 phosphatase, is a negative regulator of the PI3K/Akt pathway and thus an inducer of autophagy [112].

3.2.2. *AMP-activated protein kinase*

AMP-activated protein kinase (AMPK), a serine/threonine protein kinase, is another sensor of cellular energy status and regulates the metabolism of glucose and lipids in response to changes in nutrient and intracellular ATP concentration. In conditions of nutrient deprivation, reduced ATP production causes an elevated AMP/ATP ratios that activates the energy-sensing serine/threonine kinase 1(LKB1)-AMPK signaling axis [113]. Elevated AMP/ATP ratio activates the LKB1 and subsequently phosphorylates and activates AMPK. The activation of AMPK mediates the phosphorylation of TSC, which results in the inactivation of mTORC1 and induction of autophagy [114, 115]. Moreover, AMPK has also been found to directly phosphorylate RAPTOR, an activating component of mTORC1, thereby inhibits mTORC1 in a TSC-independent manner (Figure 2B) [116].

AMPK also directly regulates autophagy through the phosphorylation and activation of ULK1. Studies have shown that AMPK interacts with the N-terminal proline/serine (PS)-rich domain of ULK1, and this interaction is required for the ULK1-mediated autophagy (Figure 2B) [117]. Furthermore, AMPK has been shown to associate with and directly phosphorylate ULK1 on several amino acid sites, and this modification is required for ULK1 activation in response to nutrient deprivation [118]. It was reported that AMPK can interact with and phosphorylate ULK1 at Ser555, Ser637, and Thr659 and that AMPK-dependent phosphorylation of ULK1 is involved in the localization of ATG9 and increases autophagy efficiency [119]. Similarly, Kim *et al.* found that under the conditions of glucose starvation, AMPK activates ULK1 through direct phosphorylation of ULK1 on Ser317 and Ser777, thereby activating ULK1 and promoting autophagy [120]. By contrast, Shang *et al.* found that ULK1 undergoes dramatic dephosphorylation on Ser638 and Ser758 upon starvation, and the dephosphorylation of ULK1 leads to its dissociation from AMPK and becoming more active in autophagy induction [113, 121].

3.3. Regulation of autophagy by oncogenes and tumor suppressor

3.3.1. Bcl-2 family members

Bcl-2 family members were initially identified and characterized as regulators of apoptosis; however, more and more evidence has revealed that the members of this family also regulate the autophagy process. The antiapoptotic Bcl-2 family members, including Bcl-2, Bcl-xL, Bcl-w, and Mcl-1, have been found to interact with Beclin 1 and inhibit autophagy [122-124] (see Figure 2B). Because the overexpression of these antiapoptotic genes is commonly seen in cancer cells, the inhibition of autophagy may promote the oncogenic properties of these Bcl-2 family proteins. Indeed, the small interfering RNA (siRNA)-mediated knockdown of Bcl-2 can trigger autophagy and apoptosis in tumor cells [125, 126]. The antiapoptotic proteins were found to interact with Beclin 1 through their BH3 receptor domain and the BH3 domain of Belcin 1, thereby inhibiting Beclin-1-dependent autophagy [122]. The disruption of such interaction by ABT737, a BH3-mimetic agent, or the expression of other proteins with Bcl-2 homology 3 (BH3) domain that competitively disrupt the interaction can induce autophagy [123]. Additionally, the death-associated protein kinase, DAPK, a protein that phosphorylates Beclin 1 thereby disrupting Beclin-1-Bcl-2 interaction and inducing autophagy, is another inducer of autophagy that is commonly silenced by methylation in different types of human cancers.

In contrast to antiapoptotic proteins, the proapoptotic BH3-only proteins, such as BNIP3L, Bad, Noxa, Puma, BimEL22, and Bik, can promote autophagy [127]. For example, autophagy induced by hypoxia occurs through a hypoxia inducible factor-1 (HIF-1)-dependent transcriptional activation of BNIP3L that disrupts the interaction between Bcl-2 and Beclin 1 [128]. Furthermore, Puma, another "BH3-only" protein, induces mitochondrial autophagy in response to mitochondrial perturbations in a Bax/Bak-dependent manner [129].

3.3.2. p53 and p53-related regulators

The p53 tumor suppressor protein is well known for its role as a transcription factor that regulates the expression of a series of genes, contributing to cell cycle arrest, DNA damage and repair, apoptosis, and senescence [130, 131]. Underscoring its importance in the regulation of proliferative homeostasis, it is the most commonly mutated tumor suppressor in human cancer [132]. In mammalian cells, p53 shuttles between the nucleus and the cytoplasm; activated p53 translocates to the nucleus and induces the expression of target genes. Although p53 is best known as a nuclear transcription factor, studies also demonstrate that cytoplasmic p53 mediates mitochondrial outer membrane permeabilization and transcription-independent apoptosis [133, 134].

Similarly, p53 has been shown to modulate autophagy depending on its subcellular localization. Nuclear p53 stimulates autophagy in a transcription-dependent fashion by modulating the expression of a number of regulators that inhibit the mTOR pathway. For example, p53 activates the genes encoding AMPKβ1 (a component of AMPK), TSC1/2 and PTEN, which are all known negative regulators of mTORC1, leading to the activation of autophagy [135]. In addition, sestrin 1 and sestrin 2, two p53 target genes, have been identified as a critical link between p53 activation and mTORC1 activity [136]. Sestrin is induced in response to DNA damage and oxidative stress in a p53-dependent manner and inhibits mTORC1 activity via AMPK-mediated TSC activation (Figure 2B) [136]. Furthermore, sestrin 2 was shown to be required for autophagy induction in response to various cellular stress conditions, including nutrient starvation and rapamycin exposure [137]. In addition, p53 can promote autophagy in an mTOR-independent manner via the upregulation of damage regulated autophagy modulator (DRAM), a lysosomal protein mediating autophagic cell death [138]. However, p53 can also suppress autophagy [139], an effect attributable to cytoplasmic rather than nuclear p53 (Figure 2B). A recent study indicates that cytoplasmic p53 can regulate autophagy through direct interaction with FIP200 [140]. In addition, several autophagy inducers, such as nutrient starvation, rapamycin, and ER stress, stimulate proteasome-mediated degradation of p53; hence, the inhibition of p53 degradation can suppress autophagy induced by these cellular stress signals.

4. Necroptosis/programmed necrosis

For a long time, necrosis has been considered as a form of cell death that is uncontrolled and lacks underlying signaling events resulting in dramatic irreversible alterations in essential cell parameters of metabolism and cell structure [141]. This might be true for cell death in response to severe physical or chemical damage or adverse conditions; however, accumulating evidence supports the notion that necrosis is a regulated process involving multiple developmental, physiological, and pathological scenarios [142, 143]. For such a reason, it is called necroptosis or programmed necrosis. Necroptosis is characterized by cytoplasmic and organelle swelling, followed by the disruption of the cell membrane integrity, leading to the release of the cellular contents into the extracellular milieu, which may result in an inflammatory response. Unlike apoptosis, the nuclei of necrotic cells remain largely intact [144]. Necroptosis can be induced by inhibition of cellular energy production, generation of ROS, imbalance of intracellular calcium flux, or extracellular cell death signals, which are also able to induce apoptosis, suggesting that different types of cell death may share, at least in part, common mechanisms. In this sense, time and intensity of stimulus may determine the type of cell death. Indeed, one study showed that depending on glutathione depletion and oxidative stress level, apoptosis can switch to necroptosis [145].

4.1. Molecular mechanisms of necroptosis

Necroptosis can be triggered by ligands through numerous death receptors, including TNFR1 and TNFR2, TRAILR1 and TRAILR2, CD95 (Fas), and toll-like receptors (TLRs) [146-148], as well as by different kinds of physical-chemical stress stimuli, such as anticancer drugs, ionizing radiation, and calcium overload [149].

The death receptors are activated by their ligands followed by the recruitment and activation of caspase 8 that trigger the apoptosis in the absence of NF-κB survival pathway. However, under conditions that fail to trigger apoptosis, necroptosis may be an alternative cell death pathway. As shown in Figure 3, in the context of TNFR1 signaling, TNFα activates TNFR1, which in turn induces the recruitment of RIP1 kinase and other proteins, including TRADD, TRAF2, and cIAP1/2, to form a transient molecular complex referred to as complex I [144]. In complex I, RIP1 is rapidly modified by k63-linked polyubiquitination mediated by E3 ligases, cIAP1, and cIAP2. The ubiquitination of RIP1 serves as a platform to dock additional signaling molecules, IKK complex, key mediators that lead to the activation of the canonical NF-κB signaling, or inflammatory pathways [150-152]. RIP1 can be subsequently deubiquitinated by the enzyme cylindromatosis (CYLD) [153] and, together with proteins involved in cell death signaling, form complex II, which comprises as key components RIP1, TRADD, FADD, and caspase 8 [154]. The formation of complex II initiates the cell death signals, and the cell death via apoptosis or necroptosis is determined at this step. In the absence of cIAP1 or FLIP, RIP1, FADD, and caspase 8 form complex IIa to activate the caspase cascade and to induce apoptosis [31]. However, when caspase 8 activation is inhibited due to genetic or pharmacological inhibition, RIP1 together with RIP3 forms a complex that leads to the necroptotic signal transduction pathway [155]. This RIP1/RIP3-containing cytoplasmic necroptotic protein complex is called complex IIb (also known as necrosome), which constitutes a key molecular platform of necroptosis. RIP1 and RIP3 can phosphorylate reciprocally in an autocrine/ paracrine manner, leading to the activation of their kinase activity [32, 156].

The mixed lineage kinase domain-like protein (MLKL), also detected in complex IIb, is the most downstream effector of necroptosis so far identified [157, 158]. The N-terminal domain of MLKL is required for assembly of higher order structure and recruitment of MLKL to the plasma membrane, followed by permeabilization of the plasma membrane [159-161]. The C-terminal pseudokinase domain of MLKL interacts with RIP3 and is phosphorylated by the latter at the threonine 357 and serine 358 residues, and these phosphorylation events are critical for necroptosis [161]. In fact, blocking MLKL activity leads to necroptosis inhibition. Besides the plasma membrane, activated MLKL may translocate to intracellular membranes, possibly leading to the permeabilization of the ER, mitochondria and lysosome [160]. In addition, RIP3 phosphorylates MLKL, which in turn activates the mitochondrial phosphatase phosphoglycerate mutase 5 (PGAM5), a central downstream effector of the necrosomal complex. PGAM5 in turn initiates the dephosphorylation of GTPase dynamin-related protein 1 (DRP1), a mitochondrial fission regulator, which leads to the mitochondrial fission and mitochondrial fragmentation. This process is essential for the necroptotic pathways, as necrosis cannot occur without mitochondrial fission [162].

In addition to necrosome, another necroptosis-inducing complex referred to as "ripoptosome" has also been identified [163, 164]. Under normal conditions, the core components of this complex, namely, RIP1, FADD, and caspase 8, are ubiquitinated by IAPs, which leads to the degradation of these core components and thereby suppresses ripoptosome formation (Figure 3). However, when exposed to Smac mimetics or genotoxic stress, IAPs are downregulated, resulting in the spontaneous formation of the ripoptosome and triggering of caspase-8-

http://dx.doi.org/10.5772/61260

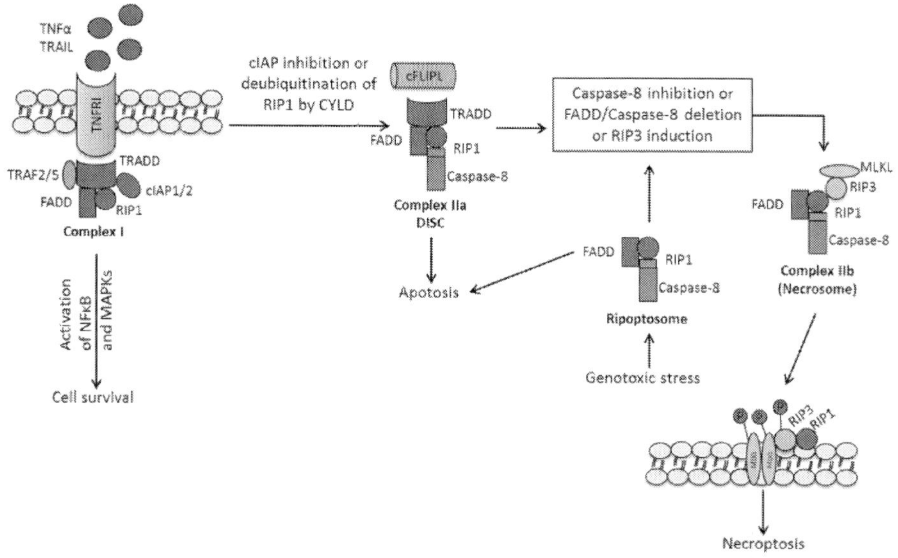

Figure 3. TNFR1 engagement- or genotoxic stress-induced formations of divergent signaling complexes, leading to the activation of NF-κB, apoptosis and necroptosis. Binding of ligands to receptors leads to the intracellular formation of complex I. Lys63-linked polyubiquitination of RIP1 (in complex I) by cIAP ligases results in cell survival through the activation of NF-κB and MAPKs. Deubiquitination of RIP1 by CYLD or inhibition of cIAP proteins leads to the conversion of complex I to complex IIa, activating the caspase cascade for apoptosis induction. Under conditions where caspase 8 activity is inhibited or RIP3 is highly expressed, RIP1 interacts with RIP3 to form complex IIb (necrosome), which mediates necroptosis. The formation of complex IIb requires the kinase activity of RIP1. RIP3 and MLKL are phosphorylated in complex IIb and translocated to the plasma membrane, where the complex mediates membrane permeabilization. In addition, the downregulation of IAPs by Smac mimetics or genotoxic stress results in the spontaneous formation of the ripoptosome, which can trigger caspase-8-mediated apoptosis or caspase-independent necroptosis.

mediated apoptosis or caspase-independent necroptosis. There are three signaling pathways initiated following ripoptosome formation. (1) The formation of caspase 8 homodimers within the complex results in full catalytic activity and thus apoptosis. (2) The formation of caspase-8-cFLIPL (long splice form of FLIP) heterodimers instead results in limited catalytic activity, which is able to cleave RIP1 but is not sufficient to trigger apoptosis, leading to ripoptosome disassembly and cell survival. (3) The formation of caspase-8-cFLIPS (short splice form of FLIP) heterodimers predominates within the complex, and caspase 8 activation and RIP1 cleavage are prevented, thereby promoting ripoptosome formation. This in turn leads to the mode of cell death instead of being switched to necroptosis [163, 164].

A schematic overview of major signal transduction pathways induced by various stimuli and ultimately leading to necroptosis can be found in the review article (Figure 1) by Kaczmarek *et al.* [165].

4.2. Regulation of necroptosis

4.2.1. FLIP and necroptosis

cFLIP molecule has been shown to be able to modulate the activation of procaspase 8 and thereby prevents the apoptosis mediated by death receptors [166]. In the cytoplasm, RIP1 can form complex with FADD, caspase 8, and TRADD (further referred to as ripoptosome) following stimulation of T-cell receptor, TLR3, or TRL4. Remarkably, in response to genotoxic stress (DNA damage), the spontaneous formation of the ripoptosome occurs independent of death receptor activation [4, 144, 156, 163]. The ripoptosome can induce caspase-dependent or caspase-independent cell death, depending on the cellular context or differential regulation of caspase 8 by cFLIP. The cleavage of RIP1 in the ripoptosome complex by caspase 8 homodimers triggers the downstream activation of effector caspases, leading to the induction of apoptosis. Further, ripoptosome-mediated cell death or necroptosis also depends on the type of FLIP isoform. The caspase 8/FLIPL heterodimers may induce RIP cleavage, thus leading to ripoptosome disassembly and necroptosis inhibition, whereas caspase 8-/cFLIPS lacks proteolytic activity necessary for RIP1 degradation, thus leading to necroptosis induction via RIP1 and RIP3 [164]. Therefore, it is possible to divert cells to undergo apoptosis via the inhibition of necroptosis through the modulation of ratio of FLIPL to FLIPS. Indeed, previous studies have demonstrated that cFLIP protects cIAP antagonist-treated cells from Fas-induced cell death, which involves both apoptosis and necroptosis [167]. cFLIPL inhibits the formation of the cell death-inducing "ripoptosome, " which functions in TLR3-induced apoptosis and necroptosis [164]. Furthermore, siRNA-mediated silencing of cFLIPL sensitizes cells to TNF-induced RIP1/RIP3-dependent necroptosis [168]. Therefore, FLIP plays a pivotal role not only in the regulation of apoptosis but also in necroptosis via the formation of ripoptosome and by switching between apoptotic and necroptotic mechanism.

4.2.2. IAPs and necroptosis

The members of the IAP protein family exhibit E3 ubiquitin ligase activity and are characterized by BIR domains that bind the active sites of caspases and inhibit proteolytic function [169]. During the intrinsic pathway of apoptosis, Smac/DIABLO is released from mitochondria to cytosol thereby releases the caspases from the trap of IAP, leading to the activation of caspases followed by apoptotic cell death. Smac protein was shown to induce the autodegradation of cytosolic IAP1 and IAP2, allowing the formation of a caspase-8-activating complex consisting of RIP1, FADD, and caspase 8 [170]. Several mammalian IAPs may utilize ubiquitination to regulate their own stability. It has been recently found that Chal-24-induced autophagy activation can result in the degradation of c-IAP1 and c-IAP2 and the formation of ripoptosome, thus contributing to necroptosis induction [171]. Remarkably, in the absence of IAPs and under conditions where caspases are blocked, necroptosis can be stimulated via RIP1 and its downstream kinase [172]. It has been demonstrated that loss of cIAPs promotes the spontaneous formation of ripoptosome induced by genotoxic stress or TLR3 stimulation through poly (I:C), a synthetic homologue of virus-derived double stranded DNA. Such event occurs independently of death receptor stimulation and is suppressed by the cIAP1 or cIAP2 that cause RIP1 ubiquitination and degradation [164]. FLIPL knockdown is able to enhance ripoptosome aggregation, thus sensitizing cells to etoposide or TLR3-mediated cell death. The role played by ripoptosome is complex since it can stimulate caspase-8-mediated apoptosis or caspase-independent necroptosis depending on the cell types [163].

4.2.3. *Regulation of necroptosis by caspase 8 activity*

The concept of apoptosis blocking necrosis by caspase activity was firstly proved in 1998 by the finding that the pharmacological inhibition of caspase activity sensitizes TNF-mediated necrotic cell death in L929 cells [173]. Since then, this concept has been accepted as an established theory. Indeed, zVAD-fmk, a pan-caspase inhibitor, has been widely used to induce necroptosis in a variety of cell lines as well as in mice models [173, 174]. Among the caspases, caspase 8 is responsible for the switching between apoptosis and necroptosis [174]. Necroptosis but not apoptosis was observed in caspase-8-deficient Jurkat cell lines in response to Fas and TNFR stimulation. *In vivo* studies have also shown that caspase-8-deficient mice have significant necroptotic death, leading to embryonic lethality, which support the roles of caspase 8 in blocking necroptosis. T-cell- or intestinal epithelial cell-specific deletion of caspase 8 in mice also exhibited severe necroptotic features, inducing immunodeficiency or terminal ileitis, respectively [175, 176].

The critical roles of caspase 8 on necroptosis are known to induce the cleavage of RIP1 and RIP3 [31, 177]. Therefore, cells treated with caspase inhibitor or deficient in caspase 8 increase the RIP1-RIP3 complex formation as well as necroptosis [156, 172, 178]. In addition, CYLD was recently identified as a target of caspase 8 and a critical mediator. Caspase 8-mediated CYLD cleavage at Asp215 prevents necroptosis, whereas the expression of mutant CYLD (D215A), which is resistant to caspase 8-mediated cleavage, enhances necrosome formation and necroptosis [178].

4.2.4. *RIP1/RIP3 and necroptosis*

RIP1 and RIP3 are key signaling molecules in inducing necrosis and are regulated by caspases and ubiquitination. The activity of RIP1 is specifically associated with necrosis and not with apoptosis, which is demonstrated by use of Necrostatin-1 (Nec-1) that specifically blocks the kinase activity of RIP1 [179]. Nec-1 inhibits TNF-induced necrosis in L929 cells and FasL-induced necrosis in Jurkat cells that were pretreated with caspase inhibitor zVADfmk or deficient in FADD [180]. In addition to RIP1, RIP3 kinase activity is also involved in caspase-independent cell death [177], and it has become clear that RIP3 determines cells to undergo necrosis in response to TNF treatment [32, 156, 172]. In contrast to RIP1, RIP3 is not required for TNF-induced NF-κB activation [156, 181]. RIP1 activity is essential for necrosome formation and its C-terminal RHIM domain (RIP homotypic interaction motif), allowing homotypic interaction with RIP3 to from a TNF-induced complex, which is important for stabilizing the necrosome [182]. Under necrotic cell death conditions, RIP3 also binds to other metabolic enzymes, such as the cytosolic glycogen phosphorylase (PYGL), the cytosolic glutamate-ammonia ligase (GLUL), and the glutaminolysis-initiating enzyme GLUD1, which positively modulates RIP3 enzymatic activity [32,

182]. These interactions result in glutamine production and regulate glycogenolysis. The knockdown of PYGL, GLUL, or GLUD1 partially reduced the degree of TNF- and zVAD-fmk-mediated ROS production and necrosis. It seems that both RIP1 and RIP3 are responsible for an increased cellular metabolism of carbohydrate and glutamine, leading to higher ROS formation and subsequent necrotic cell death [182, 183]. The activity of caspase 8 inhibits the necrotic cell death, likely by the cleavage of RIP1 and RIP3 [184], and downstream, through caspases 3 and 7 activation and poly-ADP-ribose polymerase (PARP)-1 [185]. Again, this demonstrates the importance of RIP1/3 and the enhanced ROS formation during the inhibition of caspases for the subsequent induction of necrosis [185-187].

5. Targeting cell death pathways in cancer treatment

5.1. Activating apoptosis in cancer treatment

5.1.1. TNF-Related Apoptosis-Inducing Ligand (TRAIL)

The TRAIL has been considered as a promising anticancer drug since it was found that TRAIL preferentially triggers cell death in cancer cells compared to normal cells. Furthermore, unlike TNF and FasL, TRAIL and antibodies against the TRAIL receptors were confirmed to be well tolerated and safe in nonhuman primates even at relatively high concentrations [28, 188]. Recombinant human TRAIL has been shown to have the capacity to induce apoptosis in a variety of cancer cells *in vitro* and in tumor xenografts [11]. TRAIL receptor agonists, including recombinant TRAIL, as well as humanized antibodies against TRAIL receptors have been evaluated in clinical trials [189-191]. However, several clinical trials using such drugs as single agents to induce cancer cell death did not recapitulate the promising results obtained in animal studies, which might be due to insufficient cross-linking of TRAIL receptors by the available TRAIL agonists. This has led to the investigation of TRAIL-based combination therapies to maximize antitumor activity. It has been shown that both conventional chemotherapy with DNA damaging agents and radiotherapy induce the expression of TRAIL receptors in response to DNA damage, thus suggesting a potential synergistic effect when combining these therapies with TRAIL-targeted treatment [188, 192]. For example, histone deacetylase (HDAC) inhibitors can induce the expression of TRAIL, thereby leading to the apoptosis in acute myelogenous leukemia (AML) [193]. HDAC inhibitors enhance the synthesis of several proteins involved in TRAIL signaling, such as DR5, and are able to sensitize the TRAIL-resistant cancer cells when combined with TRAIL treatment [194, 195]. In addition, enhanced assembly of the TRAIL DISC has been proposed to confer increased sensitivity in TRAIL-based combination therapies [196]. Many cytotoxic chemotherapeutic agents have been shown to induce the stabilization of p53 tumor suppressor protein in response to DNA damage and other cellular stresses, which transcriptionally activates DR5 and other proapoptotic proteins that synergizes with TRAIL. Therefore, TRAIL combination with such agents could be a useful therapeutic strategy for cancer.

However, the efficiency of TRAIL-based therapy in human cancers is not satisfactory due to the existence of both agonistic receptors (TRAIL-R1 and TRAIL-R2) and antagonistic decoy receptors (TRAIL-R3 and TRAIL-R4) in human cells. This implies that recombinant TRAIL ligand is capable of eliciting proapoptotic or antiapoptotic signals depending on the availability of these different receptors on the cell surface. Moreover, in some cancer cells, TRAIL can induce the activation of NF-κB, thus promoting cancer cell survival rather than apoptosis [188, 197]. Thus, the context-based effect of TRAIL signaling may explain the lack of efficacy seen in recent studies of TRAIL-targeted anticancer therapy. It was reported that TRADD is a key component that activates NF-κB in TRAIL signaling, and siRNA-mediated knockdown of TRADD in cancer cells sensitizes them to TRAIL-induced apoptosis. Therefore, TRADD may serve as a target for sensitizing cancer cells to TRAIL cytotoxicity [198].

5.1.2. Bcl-2 family

Antiapoptotic proteins of the Bcl-2 family, such as Bcl-2, Bcl-xL, and Mcl-1, are promising targets for anticancer drug development because they play a crucial role in regulating apoptosis, and the overexpression of these proteins is frequently observed in a variety of tumor types. Currently, three main strategies targeting this pathway are under investigation: (1) small molecules that affect gene or protein expression, (2) silencing of the upregulated antiapoptotic proteins with antisense oligonucleotides, and (3) BH3-only peptides or synthetic small molecule inhibitors interfering with Bcl-2 like protein function. With regard to transcription silencing, studies have shown that, depending on the tissue origin of the malignancy, the expression of Bcl-2 or Bcl-xL can be downregulated in specific types of cancer and leukemia cells by small molecule drugs that modulate the activity of retinoic acid receptors (RAR), retinoid X receptors (RXR), peroxisome proliferator-activated receptors (PPAR), vitamin D receptors (VDR), and certain other members of the steroid/retinoid superfamily of ligand-activated transcription factors (SRTFs). Consequently, RAR and RXR ligands as well as PPAR modulators have been developed and evaluated for the treatment of some types of leukemia, lymphoma, and solid tumors, such as breast and prostate cancers [23, 199]. HDAC inhibitors, which function as transcriptional repressors via interaction with retinoid receptors and other transcription factors, can also favorably modulate the expression of Bcl-2 or Bcl-xL in some tumor lines [23]. These findings provide the basis for developing novel strategies for cancer treatment by suppressing the expression of antiapoptotic Bcl-2-family genes in cancer.

Besides the chemical compounds, antisense oligonucleotides have also been studied to knockdown the Bcl-2 family of antiapoptotic proteins. One agent that is currently most advanced in clinical trials is Genasense (also known as oblimersen or G3139), which is a synthetic, 18-base, single-stranded phosphorothioate oligonucleotide targeting Bcl-2 mRNA that was developed by Genta Inc. (Berkeley Heights, NJ). More precisely, Genasense is in phase II and phase III clinical trials treating a wide variety of adult and childhood tumors [200]. In addition, treatment with Genasense markedly improved the antitumor activity of many chemotherapeutic agents, such as taxanes, anthracyclines, alkylators, doxorubicin, or dacarbazine [201-203]. In a phase III clinical trial, Genasense in combination with dacarbazine was reported to significantly improve multiple clinical outcomes and increase overall survival in patients with advanced melanoma [203]. Furthermore, a bispecific antisense oligonucleotide selectively targeting Bcl-2 and Bcl-xL has been reported to simultaneously downregulate the expression of both Bcl-2 and Bcl-xL and enhance chemosensitivity in various cancer cells [204-206].

Intracellular stress signals can activate BH-3 only proteins to antagonize antiapoptotic Bcl-2 family members. An attempt to mimic the BH3-only action was the development of BH3 mimetic compounds containing exposed BH3 domain that occupy the BH3-binding site on Bcl-2 or Bcl-xL, abrogating their antiapoptotic functions. ABT-737 and its oral derivative, ABT-263 (also called navitoclax), are among the first promising BH3 mimetics in cancer therapy [207, 208]. Both drugs avidly bind and inhibit Bcl-2, Bcl-xL, and Bcl-w, but not Mcl-1 or A1. ATB737 has been shown to be effective as a single agent against certain lymphomas and small cell lung cancer *in vitro* and *in vivo* [207] and against non-small cell cancer [209]. ATB737 has also been reported to have synergistic cytotoxicity with conventional chemotherapeutic agents, and other targeted agents, including tyrosine kinase inhibitor, EGFR inhibitor, MEK inhibitor, and BRAF inhibitor, to reverse drug resistance and kill tumor cells [210, 211]. Initial clinical trials have demonstrated a significant antitumor activity of ATB263 as a single agent in the treatment of B-cell malignancies, especially CLL [212]. Several preclinical studies have shown promising effects of combinatorial use of ATB263 with conventional cytotoxic agents or targeted therapy in both solid tumor and hematologic malignancy models [210, 213]. However, the practical use of ATB263 is limited due to its propensity to induce acute thrombocytopenia. ABT-199, a newer BH3 mimetic that specifically targets Bcl-2, has been shown to suppress the growth of Bcl-2-dependent tumors *in vitro* and *in vivo* without causing thrombocytopenia since it does not antagonize Bcl-xL, which is critical for platelet survival [214-216]. Also, ATB263 has been demonstrated to enhance the antitumor activity when administrated in combination with other chemotherapeutic agents [217, 218]. Other BH3 mimetics, such as WEHI-539, BXI-61, BXI-69, Obatoclax, S1, JY-1-106, Gossypol, and its derivatives (apogossypol, apogossypolone, and TW-37) as well as selective Mcl-1 inhibitors, have been developed, and their antitumor effects have also been investigated or are under investigation (reviewed by Vogler [211]).

5.1.3. IAP inhibitors

IAPs play a critical role in the control of cell survival and death by regulating key signaling events such as caspase activation and NF-κB signaling that makes them become attractive molecular targets. XIAP has been reported to be the most potent inhibitor of apoptosis among all IAPs, consequently targeting XIAP using antisense oligonucleotides, or siRNA molecules have been developed in the treatment of cancer. Indeed, targeting XIAP by antisense oligonucleotides or siRNA has been demonstrated to be able to induce apoptosis and sensitize cancer cells to death receptor- and chemotherapeutic agents-induced cell death in a variety of cancer *in vitro* and *in vivo* [219-223]. Similarly, the siRNA-mediated downregulation of other IAPs, such as cFLIP and survivin, has also been shown to enhance chemotherapy activity in a range of cancers [224-226]. In addition, some chemical compounds, for example, mTOR inhibitors and HDAC inhibitors, can suppress the cFLIP expression via blocking its translation and transcription, respectively [226, 227].

http://dx.doi.org/10.5772/61260

Another approach for targeting IAPs is to disrupt IAP binding to caspases by small molecule IAP antagonists. IAP antagonists bind to the BIR2 or BIR3 domain of XIAP, cIAP1, and cIAP2, leading to the activation of caspase and induction of apoptosis [228]. Most of IAP antagonists are Smac mimics, and in addition to monovalent compounds that contain one Smac-mimicking unit, bivalent or dimeric IAPs have also been developed, which consists of two Smac-mimicking units that are connected via a chemical linker [229]. When used as a single agent, IAP antagonists can only effectively trigger cell death in a small subset of human malignancies, suggesting that IAP antagonist-based combination therapies might be required for the effective treatment of a majority of tumors. A variety of chemotherapeutic agents (including doxorubicin, etoposide, gemcitabine, paclitaxel, cisplatin, vinorelbine, SN38, 5-fluorouracil (5-FU), cytarabine, and HADC inhibitor vorinostat), death receptor agonists, and signal transduction modulators (including proteasome inhibitors, various kind of kinase inhibitors and monoclonal antibodies targeting growth factor receptor) have been shown to act cooperatively with IAP antagonists to enhance antitumor activity *in vitro* and in preclinical models of cancers [230-236]. For instance, beneficial synergistic effects were observed when IAP antagonists were used in combination with other compounds, such as bortezomib, TRAIL, or DNA damaging agents, such as melphalan, to reduce tumor burden in multiple myeloma models [237]. Along these lines, LBW242 was also highly beneficial in an FLT3-mutated AML xenograft mouse model when administered along with the protein kinase inhibitor PKC412 [238]. A recent study depicted the combinatorial effect of Pak1, a downstream Rac effector, inhibition on IAP antagonist treatment in NSCLC cell lines, rendering these cells hypersensitive to apoptotic cell death [239]. The development of combination therapy is warranted as it promotes better patient survival, as shown in a metastatic breast cancer Phase III clinical trial [240]. Combination therapy might promote synergistic effects leading to low drug dosage, as well as suppressing resistance to therapy if multiple cell survival pathways are targeted at once, although the probability of toxicity is also increased [241]. Fortunately, clinical trials with IAP antagonists have not showed any dose-limiting toxicity [242].

5.2. Targeting autophagy in cancer treatment

Therapeutic targeting of the autophagy pathway as a new anticancer strategy has been under extensive investigation. Since autophagy can play roles in tumor growth depending on the context, such as tumor type or stage, both of the autophagy-enhancing and autophagy-inhibiting agents may elicit beneficial effects in the treatment of cancer.

5.2.1. Inhibition of autophagy

High levels of autophagy are commonly observed in tumor cells following anticancer therapy. For example, chemotherapeutic agents (e.g., doxorubicin, temozolomide, camptothecin, and tamoxifen), HDAC inhibitors (e.g., SAHA), tyrosine kinase inhibitor (e.g., imatinib, sorafenib), and monoclonal antibody (e.g., trastuzumab) have all been demonstrated to induce autophagy in a variety of tumor cells [243-248]. Furthermore, a number of studies have shown that genetic knockdown of ATGs or pharmacological inhibition of autophagy can effectively promote cell death induced by various anticancer agents in many cancer lines and in multiple tumor models [249-252]. These findings suggest that the activation of autophagy is a protective strategy for tumor cells to avoid being entirely killed by anticancer agents. The prosurvival ability of autophagy renders tumor cells resistant to anticancer agents, which greatly compromises curative efficacy of chemotherapy. In these contexts, the inhibition of autophagy can be a promising strategy to reestablish or increase the sensitivity of tumor cells to therapeutic agents.

The common inhibitors of autophagy can be categorized into three types according to their action mechanisms: (1) inhibit the formation of autophagosome via restraining the recruitment of Class III PI3K to the membrane, such as 3-methyladenine (3-MA) and Wortmannin; (2) prevent the degradation of proteins within autophagosome by disrupting lysosomal function, such as chloroquine (CQ) and its analog hydroxychloroquine (HCQ); and (3) intervene in the fusion of autophagosome with lysosome, such as bafilomycin A1 (BafA), a direct inhibitor of vacuolar ATPase [253].

CQ and its derivative HCQ are the most common autophagy inhibitors used in clinical trials. Preclinical studies have shown that CQ and HCQ are equipotent at autophagy inhibition and potentiate the anticancer effects of different drugs both *in vitro* and *in vivo*. For example, Amaravadi *et al.* reported that in a Myc-induced lymphoma mouse model, QC and HCQ significantly enhance the cytotoxic effects of p53 expression and alkylating agents and substantially impair the recurrence of tumor after chemotherapy [254, 255]. In chronic myelocytic leukemia (CML) cell lines, the inhibition of autophagy by CQ markedly augments the cell death induced by imatinib, a tyrosine kinase inhibitor that is a first-line therapeutic agent for BCR/ABL-positive CML [256]. CQ has also been shown to promote the cytotoxic effects of SAHA, an HDAC inhibitor, to overcome imatinib-resistant CML cells [246]. In a colon cancer xenograft model, CQ in combination with vorinostat was shown to significantly reduce tumor burden and increase apoptosis [257]. Similarly, CQ enhances the anticancer effect of the saracatinib, an src inhibitor, in a xenograft mouse model of prostate cancer [258]. Currently, phase I/II clinical trials are ongoing to evaluate the potential benefits of CQ and HCQ in combination with standard cancer therapies for a variety of cancers [259], and these clinical trials are listed at http://www.clinicaltrial.gov/.

In addition to CQ and its derivatives, other potential autophagy inhibitors have also been studied for their anticancer efficacy *in vivo* and *in vitro*, including 3-MA, BafA, monensin, and pepstatin A [260-262]. For example, the inhibition of autophagy by 3-MA increases cell death induced by 5-fluorouracil (5-FU) in colon cancer xenograft model [263] as well as enhances cytotoxicity induced by imatinib in glioma cell lines [264]. BafA in combination with tyrosine kinase inhibitors, such as imatinib, nilotinib, or dasatinib, significantly increase cell death in CML cells [256]. However, it must be remembered that the cytotoxic effects of these different agents might not be solely due to the inhibition of autophagy; targeting key autophagy proteins would be a more potent and specific approach. These include ULK1, Beclin 1, or ATG proteins.

5.2.2. Induction of autophagy

Although the concept of "autophagic cell death" in mammalian cells remains largely controversial, studies do show that autophagy is required for the efficient killing of tumor cells in certain circumstances [252]. Certain tumor cells become highly resistant to apoptosis and chemotherapy by overexpression of Bcl-2 or Bcl-xL, lack of Bax and Bak, loss of Beclin 1, or exposure to pan-caspase inhibitors [76]. Most anticancer drugs exhibit limited effect on this subset of tumor cells. Fortunately, studies have demonstrated that the induction of autophagy may be an alternative way for cancer treatment when apoptosis is blocked [265, 266]; however, the conditions under which autophagy can function as a primary cell-death mechanism remain to be defined.

Among the potential targets in autophagy, Akt-mTOR pathway is the most investigated one. mTOR inhibitors, including rapamycin and its analogs everolimus (RAD-001), temsirolimus (CCI-779), and deforolimus (AP-23573), have been developed and studied for their ability to induce autophagy and cell death. Everolimus and temsirolimus have been approved for the treatment of renal cell carcinoma and mantle cell lymphoma [267]. Rapamycin has been shown to inhibit cell growth and initiate cell death in mantle cell lymphoma cell lines and various primary tumor cells, such as malignant gliomas, breast cancers, renal cell carcinomas, non-small cell lung cancers, and cervical and uterine cancers [13, 253]. Everolimus induces massive autophagy in leukemia [268], in advanced pancreatic cancer [269], and in many other cancers [270], accompanied by reduced tumor burden. In addition, everolimus in combination with etoposide, cisplatin, or doxorubicin display synergistic effects without significant increase in toxicity [271]. However, rapamycin and its analogs would inevitably activate Akt kinases, which associate with the induction of insulin receptor substrate-1, jeopardizing the antitumor effects of these mTOR inhibitors [272]. Other inhibitors, including ATP-competitive inhibitors of both mTORC1 and mTORC2 as well as the dual PI3K-mTOR inhibitor NVP-BEZ235 [273, 274], have exhibited more potent capacity to induce autophagy in cancer cells [275, 276].

Antiapoptotic Bcl-2 family members are frequently overexpressed in many human tumor types, rendering tumor cells resistant to apoptosis. Bcl-2 family members are important regulators involved in both apoptosis and autophagy. As a result, the modulation of Bcl-2 family proteins leads to not only apoptotic but also autophagic cell death. The underlying mechanism of this effect reflects the fact that Bcl-2/ Bcl-xL proteins can bind and disrupt the autophagic function of Beclin 1, which contains a BH3 domain [277]. This is notably the case for BH3 mimetics (ABT737, ABT236, gossypol, obatoclax) that targets Bcl-2/Bcl-xL, thus allowing Beclin 1 to be released to trigger autophagy [277, 278]. Obatoclax has been shown not only to induce cell death on its own but also to potentiate the effects of other anticancer agents such as the dual EGFR/HER2 inhibitor lapatinib, or HDAC inhibitors [279]. Although the inhibitory effect affects both apoptosis and autophagy, the tumor cells preferentially undergo autophagic cell death in apoptosis-defective cells. For example, doxorubicin primarily induces autophagy at low doses while apoptosis at high doses, and the combination of Bcl-2 siRNA treatment with a low dose of doxorubicin enhances the autophagic response, tumor growth inhibition, and cell death [126].

In pursuit of new drugs to selectively kill renal cell carcinoma (RCC), Giaccia and colleagues identified a compound, STF-62247, that strongly induced autophagy and massive vacuolization in VHL (a tumor suppressor gene lost in 75% of RCCs)-deficient RCC cells with no apparent apoptosis induction. Blocking autophagy using ATG5 or ATG7 siRNA or 3-MA prevents STF-62247-induced cell death, indicating that this compound induces cell death by autophagy in VHL-deficient RCC cells [280]. In addition, other autophagic cell death-inducing anticancer agents have been developed and studied [279].

5.3. Modulation of necrosis for cancer treatment

The accumulating data have indicated that necrotic cell death can be activated to induce the damage of tumor tissue. DNA-damaging agents are the most widely used and effective chemotherapeutic approach for cancer treatment, which has been shown to be able to stimulate a regulated form of necrosis [281]. The PARP is activated in response to DNA damage, which facilitates the access of DNA repair enzymes to damaged DNA. The hyperactivation of PARP depletes cytosolic nicotinamide adenine dinucleotide (NAD) and induce necrosis [282], which may lead to selectively killing tumor cells because highly proliferating tumor cells depend on cellular NAD to generate energy through aerobic glycolysis. PARP-mediated necrosis may explain the phenomenon that neither Bax/Bak nor p53 deficiency impedes cell death in response to DNA-damaging agents. Jouan-Lanhouet *et al.* found that the death receptor ligand TRAIL induces necroptosis in human HT29 colon and HepG2 liver cancer cells via RIP1/RIP3-dependent PARP1 activation and depletion of cellular ATP levels, suggesting PARP1 activation as an effector mechanism downstream of RIP1/RIP3 [283].

There are more and more compounds and anticancer drugs that have been demonstrated to induce cancer cell death through necrosis, such as shikonin, FTY720, staurosporine, derivatives of amiloride (5-benzylglycinyl-amiloride and glycinyl-amiloride), and BI2536 (a small molecule inhibitor of the mitotic kinase Plk1). Most of them were not necessarily designed in a mechanism-based fashion but were only later found to induce necrotic features in the dying cells [284-289]. Shikonin, a naturally occurring naphthoquinone, was reported to induce necroptotic cell death in cancer cells that can be prevented by RIP1 inhibitor necrostatin-1, a specific inhibitor of necroptosis. Moreover, shikonin-induced necroptosis can overcome drug- and apoptosis-resistant cancer cell lines overexpressing P-glycoprotein, MRP1, BCRP, Bcl-2, or Bcl-xL [284, 285]. FTY720, a sphingolipid analogue drug that mimics ceramide, was shown to target the I2PP2A/SET oncoprotein, which results in the activation of tumor suppressor PP2A and subsequently induces RIP1-mediated necroptotic cell death and tumor growth inhibition [286]. Staurosporine, an inhibitor of a broad spectrum of protein kinases, has been shown to induce necroptosis in leukemia cells when caspase activation is inhibited. The induction of necroptosis was blocked by several pharmacological inhibitors, including necrostatin-1, HSP90 inhibitor geldanamycin, MLKL inhibitor necrosulfonamide, and a cathepsin inhibitor CA-074-OMe, which has been demonstrated to rescue the caspase-independent necrotic cell death of leukemia cells treated by staurosporine [289]. Other anticancer agents have also been shown to induce necroptosis via different mechanisms.

Bonapace *et al.* reported that obatoclax (GX15-070), a putative antagonist of Bcl-2 family members, could overcome glucocorticoid resistance in childhood acute lymphoblastic leukemia (ALL) through the induction of autophagy-dependent necroptosis, which bypassed the block in mitochondrial apoptosis [290]. Obatoclax was also shown to promote the assembly of the necrosome on autophagosomal membranes, thereby connecting obatoclax-induced autophagy to necroptosis signaling pathways [291]. Coimmunoprecipitation assays demonstrated that obatoclax promoted the physical interaction of ATG5, a constituent of autophagosomal membranes, with FADD, RIP1, and RIP3 as key components of the necrosome [291].

Small molecule inhibitors targeting IAPs such as Smac mimetics, which have been developed to induce apoptotic cancer cell death, have been found to also engage necroptotic cell death. He *et al.* first reported that, upon the inhibition of caspase activity, Smac mimetics in combination with TNFα provoke a strong necroptotic response in Smac mimetic-resistant cancer cells [172]. Similarly, Smac mimetic BV6 promotes TNFα-induced necroptosis not only in leukemia cell lines deficient in caspase 8 or FADD but also in primary, patient-derived ALL cells [292]. This Smac mimetic/TNFα-triggered necroptosis occurred in an RIP1-dependent but caspase-independent manner in these leukemia cells lacking caspase 8 or FADD; however, in FADD- or caspase-8-proficient leukemia cells, the same cotreatment of Smac mimetic and TNFα induced apoptotic cell death [292]. This illustrates that Smac mimetic can prime leukemia cells to TNFα-mediated cell death via either necroptosis or apoptosis depending on the cellular context. Mechanistically, by promoting the degradation of cIAP proteins, Smac mimetics stimulate the necrosome formation and promote necroptosis [293]. Furthermore, a recent study demonstrated that Smac mimetic cooperates with demethylating agents to synergistically induce cell death and can circumvent apoptosis resistance of AML cells by switching to necroptosis [294].

Targeted toxins are fusion proteins that combine a targeting protein, such as a ligand for a specific receptor, and a toxic peptide derived from a bacterial pathogen [295]. Diphtheria toxin GM-CSF (DT-GMCSF) was shown to kill AML cells by simultaneously activating both caspase-dependent apoptosis and caspase-independent necroptosis [296]. Interestingly, DT-GMCSF-induced necroptotic cell death even occurred in apoptosis-resistant AML cells, indicating that necroptosis may open new perspectives for cancer drug development in AML [296].

6. Conclusions and future perspectives

The dysregulation of the cell death process is closely related to cancer progression and resistance to chemotherapy. The concept to therapeutically target apoptotic signal transduction pathways has significant implications for cancer therapy since intact apoptosis programs are critically required for the antitumor activity of most current cancer therapies that are used in clinical oncology. The reactivation of apoptosis not only directly induces cell death in tumor cells but also sensitizes tumor cells to other chemotherapeutic or targeted therapy agents. However, given the wide variety of genetic and epigenetic defects that lead to apoptosis resistance in most cancers, understanding the mechanism and regulation of other cell death pathways in response to antitumor agents is important. Autophagy and necrosis are two nonapoptotic cell death models that can be triggered as a consequence of cancer therapy and may have overlapping but separable regulatory networks. Unlike apoptosis or necrosis, autophagy has been shown to exert dual functions in cancers. On one hand, autophagy can function as a survival pathway in response to anticancer agents; hence, autophagy inhibitors may be used as adjuvants to standard cancer therapies. On the other hand, autophagy can lead to cell death in certain circumstances; thus, autophagy inducers may help to eradicate cancer cells. As the effect of autophagy on cancer therapy varies depending on cell context such as cell type, phases, and microenvironment, personalized pharmacotherapeutic strategies should be adopted alone or most likely in combination with standard chemotherapeutic agents.

The success of compounds like the BH3 mimetics, Smac mimetics, TRAIL, and mTOR inhibitors in preclinical studies is proof that the reactivation of defective cell death pathways in cancer cells is possible and can effectively eradicate tumor cells. Therefore, efforts should be put to further delineate and identify regulatory components of the different cell death pathways that can be targeted and manipulated to execute death. Understanding the crosstalk between the different cell death pathways as well as between cell death and non-cell death pathways in cancer is crucial to spot prospective convergence points between pathways. Targeting the convergence points will allow the switching between pathways and improvisation of alternate means of inducing cancer cell death. For example, TNF signaling to the NF-κB prosurvival and inflammatory pathway can be rerouted to the necroptotic pathway to promote cell death just by modulating cIAP levels [163].

Author details

Yong Zhong Xu[1], Cynthia Kanagaratham[1], Mina Youssef[1] and Danuta Radzioch[1,2*]

*Address all correspondence to: danuta.radzioch@mcgill.ca

1 Department of Human Genetics, McGill University, Montreal, Quebec, Canada

2 Department of Medicine, McGill University, Montreal, Quebec, Canada

References

[1] Ferlay J, Soerjomataram I, Ervik M, Dikshit R, Eser S, Mathers C, et al. GLOBOCAN 2012 v1. 0, Cancer Incidence and Mortality Worldwide: IARC CancerBase No. 11. Lyon, France: International Agency for Research on Cancer; 2013. Visit http//globocan iarc fr. 2014.

[2] Faber AC, Ebi H, Costa C, Engelman JA. Apoptosis in targeted therapy responses. the role of BIM. Adv Pharmacol. 2012;65(1):519-42.

[3] Guo M, Hay BA. Cell proliferation and apoptosis. Curr Opin Cell Biol. 1999;11(6): 745-52.

[4] Long JS, Ryan KM. New frontiers in promoting tumour cell death: targeting apoptosis, necroptosis and autophagy. Oncogene. 2012;31(49):5045-60.

[5] Fuchs Y, Steller H. Programmed cell death in animal development and disease. Cell. 2011;147(4):742-58.

[6] Tan ML, Ooi JP, Ismail N, Moad AIH, Muhammad TST. Programmed cell death pathways and current antitumor targets. Pharm Res. 2009;26(7):1547-60.

[7] Bialik S, Zalckvar E, Ber Y, Rubinstein AD, Kimchi A. Systems biology analysis of programmed cell death. Trends Biochem Sci. 2010;35(10):556-64.

[8] Singh R, George J, Shukla Y. Role of senescence and mitotic catastrophe in cancer therapy. Cell Div. 2010;5(4).

[9] Dimri GP. What has senescence got to do with cancer? Cancer Cell. 2005;7(6):505-12.

[10] Debatin K-M. Apoptosis pathways in cancer and cancer therapy. Cancer Immunol Immunother. 2004;53(3):153-9.

[11] Ashkenazi A. Targeting death and decoy receptors of the tumour-necrosis factor superfamily. Nat Rev Cancer. 2002;2(6):420-30.

[12] Hanahan D, Weinberg R a, Francisco S. The Hallmarks of Cancer Review University of California at San Francisco. Horm Res. 2000;100(1):57-70.

[13] Kondo Y, Kanzawa T, Sawaya R, Kondo S. The role of autophagy in cancer development and response to therapy. Nat Rev Cancer. 2005;5(9):726-34.

[14] Qu X, Yu J, Bhagat G, Furuya N, Hibshoosh H, Troxel A, et al. Promotion of tumorigenesis by heterozygous disruption of the beclin 1 autophagy gene. J Clin Invest. 2003;112(12):1809-20.

[15] Kimmelman AC. The dynamic nature of autophagy in cancer. Genes Dev. 2011;25(19):1999-2010.

[16] Wu W, Liu P, Li J. Necroptosis: an emerging form of programmed cell death. Crit Rev Oncol Hematol. 2012;82(3):249-58.

[17] McCall K. Genetic control of necrosis—another type of programmed cell death. Curr Opin Cell Biol. 2010;22(6):882-8.

[18] J. F. R. Kerr, A. H. Wyllie and ARC. Apoptosis: a basic biological phenomenon with wide-ranging implications in tissue kinetics. J Intern Med. 1972;26(4):239-57.

[19] Liu J, Lin A. Role of JNK activation in apoptosis: a double-edged sword. Cell Res. 2005;15(1):36-42.

[20] Lockshin RA, Zakeri Z. Cell death in health and disease. J Cell Mol Med. 2007;11(6): 1214-24.

[21] Kroemer G, Galluzzi L, Vandenabeele P, Abrams J, Alnemri ES, Baehrecke EH, et al. Classification of cell death: recommendations of the Nomenclature Committee on Cell Death 2009. Cell Death Differ. 2009;16(1):3-11.

[22] Kumar V, Abbas AK, Fausto N, Aster JC. Robbins and Cotran Pathologic Basis of Disease, Professional Edition: Expert Consult-Online. 8th Ed. Philadelphia: Elsevier Health Sciences; 2014. pp. 25-32.

[23] Hassan M, Watari H, Abualmaaty A, Ohba Y, Sakuragi N. Apoptosis and molecular targeting therapy in cancer. Biomed Res Int. 2014;2014(1):150845.

[24] Hanahan D, Weinberg RA. Hallmarks of cancer: the next generation. Cell. 2011;144(5):646-74.

[25] Parrish AB, Freel CD, Kornbluth S. Cellular mechanisms controlling caspase activation and function. Cold Spring Harb Perspect Biol. 2013;5(6):pii: a008672.

[26] Szegezdi E, Fitzgerald U, Samali A. Caspase-12 and ER-stress-mediated apoptosis: the story so far. Ann N Y Acad Sci. 2003;1010(1):186-94.

[27] Locksley RM, Killeen N, Lenardo MJ. The TNF and TNF receptor superfamilies: integrating mammalian biology. Cell. 2001;104(4):487-501.

[28] Ashkenazi A. Targeting the extrinsic apoptosis pathway in cancer. Cytokine Growth Factor Rev. 2008;19(3-4):325-31.

[29] Peter ME, Krammer PH. The CD95(APO-1/Fas) DISC and beyond. Cell Death Differ. 2003;10(1):26-35.

[30] Wilson NS, Dixit V, Ashkenazi A. Death receptor signal transducers: nodes of coordination in immune signaling networks. Nat Immunol. 2009;10(4):348-55.

[31] Lin Y, Devin A, Rodriguez Y, Liu ZG. Cleavage of the death domain kinase RIP by caspase-8 prompts TNF-induced apoptosis. Genes Dev. 1999;13(19):2514-26.

[32] Zhang D-W, Shao J, Lin J, Zhang N, Lu B-J, Lin S-C, et al. RIP3, an energy metabolism regulator that switches TNF-induced cell death from apoptosis to necrosis. Science. 2009;325(5938):332-6.

[33] Scaffidi C, Fulda S, Srinivasan A, Friesen C, Li F, Tomaselli KJ, et al. Two CD95 (APO-1/Fas) signaling pathways. EMBO J. 1998;17(6):1675-87.

[34] Schimmer AD, Hedley DW, Penn LZ, Minden MD. Receptor- and mitochondrial-mediated apoptosis in acute leukemia: a translational view. Blood. 2001;98(13): 3541-53.

[35] Roy S, Nicholson DW. Cross-talk in cell death signaling. J Exp Med. 2000;192(8):F21-6.

[36] R Safa A. Roles of c-FLIP in apoptosis, necroptosis, and autophagy. J Carcinog Mutagen. 2013;Suppl 6.:pii: 003.

[37] Thome M, Schneider P, Hofmann K, Fickenscher H, Meinl E, Neipel F, et al. Viral FLICE-inhibitory proteins (FLIPs) prevent apoptosis induced by death receptors. Nature. 1997;386(6624):517-21.

[38] Wertz IE, Dixit VM. Regulation of death receptor signaling by the ubiquitin system. Cell Death Differ. 2010;17(1):14-24.

[39] Gyrd-Hansen M, Meier P. IAPs: from caspase inhibitors to modulators of NF-kappaB, inflammation and cancer. Nat Rev Cancer. 2010;10(8):561-74.

[40] Mayo MW, Baldwin AS. The transcription factor NF-kappaB: control of oncogenesis and cancer therapy resistance. Biochim Biophys Acta. 2000;1470(2):M55-62.

[41] Jin Z, Li Y, Pitti R, Lawrence D, Pham VC, Lill JR, et al. Cullin3-based polyubiquitination and p62-dependent aggregation of caspase-8 mediate extrinsic apoptosis signaling. Cell. 2009;137(4):721-35.

[42] Karp G. Cell and molecular biology: concepts and experiments. Cell and molecular biology: concepts and experiments. 5th ed. John New Jersey: Wiley and Sons; 2008. pp. 653-7.

[43] Dewson G, Kratina T, Czabotar P, Day CL, Adams JM, Kluck RM. Bak activation for apoptosis involves oligomerization of dimers via their alpha6 helices. Mol Cell. 2009;36(4):696-703.

[44] Goldstein JC, Waterhouse NJ, Juin P, Evan GI, Green DR. The coordinate release of cytochrome c during apoptosis is rapid, complete and kinetically invariant. Nat Cell Biol. 2000;2(3):156-62.

[45] Wei MC, Zong WX, Cheng EH, Lindsten T, Panoutsakopoulou V, Ross AJ, et al. Proapoptotic BAX and BAK: a requisite gateway to mitochondrial dysfunction and death. Science. 2001;292(5517):727-30.

[46] Cory S, Adams JM. The Bcl2 family: regulators of the cellular life-or-death switch. Nat Rev Cancer. 2002;2(9):647-56.

[47] Green DR, Kroemer G. The pathophysiology of mitochondrial cell death. Science. 2004;305(5684):626-9.

[48] Bratton SB, Salvesen GS. Regulation of the Apaf-1-caspase-9 apoptosome. J Cell Sci. 2010;123(Pt 19):3209-14.

[49] Tait SWG, Green DR. Mitochondrial regulation of cell death. Cold Spring Harb Perspect Biol. 2013;5(9):pii: a008706.

[50] Kroemer G, Galluzzi L, Brenner C. Mitochondrial membrane permeabilization in cell death. Physiol Rev. 2007;87(1):99-163.

[51] Daugas E, Nochy D, Ravagnan L, Loeffler M, Susin SA, Zamzami N, et al. Apoptosis-inducing factor (AIF): a ubiquitous mitochondrial oxidoreductase involved in apoptosis. FEBS Lett. 2000;476(3):118-23.

[52] Du C, Fang M, Li Y, Li L, Wang X. Smac, a mitochondrial protein that promotes cytochrome c-dependent caspase activation by eliminating IAP inhibition. Cell. 2000;102(1):33-42.

[53] Verhagen AM, Ekert PG, Pakusch M, Silke J, Connolly LM, Reid GE, et al. Identification of DIABLO, a mammalian protein that promotes apoptosis by binding to and antagonizing IAP proteins. Cell. 2000;102(1):43-53.

[54] Chipuk JE, Green DR. How do BCL-2 proteins induce mitochondrial outer membrane permeabilization? Trends Cell Biol. 2008;18(4):157-64.

[55] Tait SWG, Green DR. Mitochondria and cell death: outer membrane permeabilization and beyond. Nat Rev Mol Cell Biol. 2010;11(9):621-32.

[56] Youle RJ, Strasser A. The BCL-2 protein family: opposing activities that mediate cell death. Nat Rev Mol Cell Biol. 2008;9(1):47-59.

[57] Lau R, Pratt MAC. The opposing roles of cellular inhibitor of apoptosis proteins in cancer. ISRN Oncol. 2012;2012(928120):1-5.

[58] Salvesen GS, Duckett CS. IAP proteins: blocking the road to death's door. Nat Rev Mol Cell Biol. 2002;3(6):401-10.

[59] Huang Y, Park YC, Rich RL, Segal D, Myszka DG, Wu H. Structural basis of caspase inhibition by XIAP: differential roles of the linker versus the BIR domain. Cell. 2001;104(5):781-90.

[60] Srinivasula SM, Hegde R, Saleh A, Datta P, Shiozaki E, Chai J, et al. A conserved XIAP-interaction motif in caspase-9 and Smac/DIABLO regulates caspase activity and apoptosis. Nature. 2001;410(6824):112-6.

[61] Huang HK, Joazeiro CAP, Bonfoco E, Kamada S, Leverson JD, Hunter T. The inhibitor of apoptosis, cIAP2, functions as a ubiquitin-protein ligase and promotes in vitro monoubiquitination of caspases 3 and 7. J Biol Chem. 2000;275(35):26661-4.

[62] Suzuki Y, Yoda T, Ruhul A, Sugiura W. Molecular cloning and characterization of the gene coding for azoreductase from Bacillus sp. OY1-2 isolated from soil. J Biol Chem. 2001;276(12):9059-65.

[63] Chen DJ, Huerta S. Smac mimetics as new cancer therapeutics. Anticancer Drugs. 2009;20(8):646-58.

[64] Okada H, Suh W-K, Jin J, Woo M, Du C, Elia A, et al. Generation and characterization of Smac/DIABLO-deficient mice. Mol Cell Biol. 2002;22(10):3509-17.

[65] Jones JM, Datta P, Srinivasula SM, Ji W, Gupta S, Zhang Z, et al. Loss of Omi mitochondrial protease activity causes the neuromuscular disorder of mnd2 mutant mice. Nature. 2003;425(6959):721-7.

[66] Martins LM, Morrison A, Klupsch K, Fedele V, Moisoi N, Teismann P, et al. Neuroprotective role of the Reaper-related serine protease HtrA2/Omi revealed by targeted deletion in mice. Mol Cell Biol. 2004;24(22):9848-62.

[67] Zhuang M, Guan S, Wang H, Burlingame AL, Wells JA. Substrates of IAP ubiquitin ligases identified with a designed orthogonal E3 ligase, the NEDDylator. Mol Cell. 2013;49(2):273-82.

[68] Yang Z, Klionsky DJ. Eaten alive: a history of macroautophagy. Nat Cell Biol. 2010;12(9):814-22.

[69] Mijaljica D, Prescott M, Devenish RJ. Microautophagy in mammalian cells: revisiting a 40-year-old conundrum. Autophagy. 2011;7(7):673-82.

[70] Majeski AE, Fred Dice J. Mechanisms of chaperone-mediated autophagy. Int J Biochem Cell Biol. 2004;36(12):2435-44.

[71] Yorimitsu T, Klionsky DJ. Autophagy: molecular machinery for self-eating. Cell Death Differ. 2005;12(Suppl 2):1542-52.

[72] Levine B, Kroemer G. Autophagy in the pathogenesis of disease. Cell. 2008;132(1): 27-42.

[73] Turcotte S, Giaccia AJ. Targeting cancer cells through autophagy for anticancer therapy. Curr Opin Cell Biol. 2010;22(2):246-51.

[74] White E, DiPaola RS. The double-edged sword of autophagy modulation in cancer. Clin Cancer Res. 2009;15(17):5308-16.

[75] Amaravadi RK, Lippincott-Schwartz J, Yin XM, Weiss WA, Takebe N, Timmer W, et al. Principles and current strategies for targeting autophagy for cancer treatment. Clin Cancer Res. 2011;17(4):654-66.

[76] Shimizu S, Kanaseki T, Mizushima N, Mizuta T, Arakawa-Kobayashi S, Thompson CB, et al. Role of Bcl-2 family proteins in a non-apoptotic programmed cell death dependent on autophagy genes. Nat Cell Biol. 2004;6(12):1221-8.

[77] Levine B, Mizushima N, Virgin HW. Autophagy in immunity and inflammation. Nature. 2011;469(7330):323-35.

[78] Hayashi-Nishino M, Fujita N, Noda T, Yamaguchi A, Yoshimori T, Yamamoto A. A subdomain of the endoplasmic reticulum forms a cradle for autophagosome formation. Nat Cell Biol. 2009;11(12):1433-7.

[79] Hayashi-Nishino M, Fujita N, Noda T, Yamaguchi A, Yoshimori T, Yamamoto A. Electron tomography reveals the endoplasmic reticulum as a membrane source for autophagosome formation. Autophagy. 2010;6(2):301-3.

[80] Tooze SA, Yoshimori T. The origin of the autophagosomal membrane. Nat Cell Biol. 2010;12(9):831-5.

[81] Burman C, Ktistakis NT. Autophagosome formation in mammalian cells. Semin Immunopathol. 2010;32(4):397-413.

[82] Mizushima N, Ohsumi Y, Yoshimori T. Autophagosome formation in mammalian cells. Cell Struct Funct. 2002;27(6):421-9.

[83] Rabinowitz JD, White E. Autophagy and metabolism. Science. 2010;330(6009):1344-8.

[84] Efeyan A, Zoncu R, Sabatini DM. Amino acids and mTORC1: from lysosomes to disease. Trends Mol Med. 2012;18(9):524-33.

[85] Chan EYW, Kir S, Tooze SA. siRNA screening of the kinome identifies ULK1 as a multidomain modulator of autophagy. J Biol Chem. 2007;282(35):25464-74.

[86] Kihara A, Kabeya Y, Ohsumi Y, Yoshimori T. Beclin-phosphatidylinositol 3-kinase complex functions at the trans-Golgi network. EMBO Rep. 2001;2(4):330-5.

[87] Petiot A, Ogier-Denis E, Blommaart EFC, Meijer AJ, Codogno P. Distinct classes of phosphatidylinositol 3-kinases are involved in signaling pa7thways that control macroautophagy in HT-29 cells. J Biol Chem. 2000;275(2):992-8.

[88] Young ARJ, Chan EYW, Hu XW, Köchl R, Crawshaw SG, High S, et al. Starvation and ULK1-dependent cycling of mammalian Atg9 between the TGN and endosomes. J Cell Sci. 2006;119(Pt 18):3888-900.

[89] O'Farrell F, Rusten TE, Stenmark H. Phosphoinositide 3-kinases as accelerators and brakes of autophagy. FEBS J. 2013;280(24):6322-37.

[90] Funderburk SF, Wang QJ, Yue Z. The Beclin 1-VPS34 complex—at the crossroads of autophagy and beyond. Trends Cell Biol. 2010;20(6):355-62.

[91] Zhong Y, Wang QJ, Li X, Yan Y, Backer JM, Chait BT, et al. Distinct regulation of autophagic activity by Atg14L and Rubicon associated with Beclin 1-phosphatidylinositol-3-kinase complex. Nat Cell Biol. 2009;11(4):468-76.

[92] Polson HEJ, De Lartigue J, Rigden DJ, Reedijk M, Urbé S, Clague MJ, et al. Mammalian Atg18 (WIPI2) localizes to omegasome-anchored phagophores and positively regulates LC3 lipidation. Autophagy. 2010;6(4):506-22.

[93] Proikas-Cezanne T, Robenek H. Freeze-fracture replica immunolabelling reveals human WIPI-1 and WIPI-2 as membrane proteins of autophagosomes. J Cell Mol Med. 2011;15(9):2007-10.

[94] Klionsky DJ, Schulman BA. Dynamic regulation of macroautophagy by distinctive ubiquitin-like proteins. Nat Struct Mol Biol. 2014;21(4):336-45.

[95] Nishida Y, Arakawa S, Fujitani K, Yamaguchi H, Mizuta T, Kanaseki T, et al. Discovery of Atg5/Atg7-independent alternative macroautophagy. Nature. 2009;461(7264):654-8.

[96] Nakatogawa H. Two ubiquitin-like conjugation systems that mediate membrane formation during autophagy. Essays Biochem. 2013;55(1):39-50.

[97] Pankiv S, Clausen TH, Lamark T, Brech A, Bruun JA, Outzen H, et al. p62/SQSTM1 binds directly to Atg8/LC3 to facilitate degradation of ubiquitinated protein aggregates by autophagy. J Biol Chem. 2007;282(33):24131-45.

[98] Schweers RL, Zhang J, Randall MS, Loyd MR, Li W, Dorsey FC, et al. NIX is required for programmed mitochondrial clearance during reticulocyte maturation. Proc Natl Acad Sci U S A. 2007;104(49):19500-5.

[99] Sandoval H, Thiagarajan P, Dasgupta SK, Schumacher A, Prchal JT, Chen M, et al. Essential role for Nix in autophagic maturation of erythroid cells. Nature. 2008;454(7201):232-5.

[100] Eskelinen E-L, Illert AL, Tanaka Y, Schwarzmann G, Blanz J, Von Figura K, et al. Role of LAMP-2 in lysosome biogenesis and autophagy. Mol Biol Cell. 2002;13(9):3355-68.

[101] Ao X, Zou L, Wu Y. Regulation of autophagy by the Rab GTPase network. Cell Death Differ. 2014;21(3):348-58.

[102] Nara A, Mizushima N, Yamamoto A, Kabeya Y, Ohsumi Y, Yoshimori T. SKD1 AAA ATPase-dependent endosomal transport is involved in autolysosome formation. Cell Struct Funct. 2002;27(1):29-37.

[103] Jung CH, Ro SH, Cao J, Otto NM, Kim Do-Hyung DH. MTOR regulation of autophagy. FEBS Lett. 2010;584(7):1287-95.

[104] Zoncu R, Efeyan A, Sabatini DM. mTOR: from growth signal integration to cancer, diabetes and ageing. Nat Rev Mol Cell Biol. 2011;12(1):21-35.

[105] Hsu PP, Kang SA, Rameseder J, Zhang Y, Ottina KA, Lim D, et al. The mTOR-regulated phosphoproteome reveals a mechanism of mTORC1-mediated inhibition of growth factor signaling. Science. 2011;332(6035):1317-22.

[106] Efeyan A, Zoncu R, Chang S, Gumper I, Snitkin H, Wolfson RL, et al. Regulation of mTORC1 by the Rag GTPases is necessary for neonatal autophagy and survival. Nature. 2013;493(7434):679-83.

[107] Neufeld TP. TOR-dependent control of autophagy: biting the hand that feeds. Curr Opin Cell Biol. 2010;22(2):157-68.

[108] Hosokawa N, Hara T, Kaizuka T, Kishi C, Takamura A, Miura Y, et al. Nutrient-dependent mTORC1 association with the ULK1-Atg13-FIP200 complex required for autophagy. Mol Biol Cell. 2009;20(7):1981-91.

[109] Guertin DA, Sabatini DM. Defining the role of mTOR in cancer. Cancer Cell. 2007;12(1):9-22.

[110] Tee AR, Manning BD, Roux PP, Cantley LC, Blenis J. Tuberous sclerosis complex gene products, Tuberin and Hamartin, control mTOR signaling by acting as a GTPase-activating protein complex toward Rheb. Curr Biol. 2003;13(15):1259-68.

[111] Proud CG. Cell signaling. mTOR, unleashed. Science. 2007;318(5852):926-7.

[112] Shaw RJ, Cantley LC. Ras, PI(3)K and mTOR signalling controls tumour cell growth. Nature. 2006;441(7092):424-30.

[113] Shang L, Chen S, Du F, Li S, Zhao L, Wang X. Nutrient starvation elicits an acute autophagic response mediated by Ulk1 dephosphorylation and its subsequent dissociation from AMPK. Proc Natl Acad Sci U S A. 2011;108(12):4788-93.

[114] Corradetti MN, Inoki K, Bardeesy N, DePinho RA, Guan KL. Regulation of the TSC pathway by LKB1: evidence of a molecular link between tuberous sclerosis complex and Peutz-Jeghers syndrome. Genes Dev. 2004;18(13):1533-8.

[115] Inoki K, Ouyang H, Zhu T, Lindvall C, Wang Y, Zhang X, et al. TSC2 integrates Wnt and energy signals via a coordinated phosphorylation by AMPK and GSK3 to regulate cell growth. Cell. 2006;126(5):955-68.

[116] Gwinn DM, Shackelford DB, Egan DF, Mihaylova MM, Mery A, Vasquez DS, et al. AMPK phosphorylation of raptor mediates a metabolic checkpoint. Mol Cell. 2008;30(2):214-26.

[117] Herrero-Martín G, Høyer-Hansen M, García-García C, Fumarola C, Farkas T, López-Rivas A, et al. TAK1 activates AMPK-dependent cytoprotective autophagy in TRAIL-treated epithelial cells. EMBO J. 2009;28(6):1532.

[118] Egan DF, Shackelford DB, Mihaylova MM, Gelino S, Kohnz RA, Mair W, et al. Phosphorylation of ULK1 (hATG1) by AMP-activated protein kinase connects energy sensing to mitophagy. Science. 2011;331(6016):456-61.

[119] Mack HID, Zheng B, Asara JM, Thomas SM. AMPK-dependent phosphorylation of ULK1 regulates ATG9 localization. Autophagy. 2012;8(8):1197-214.

[120] Kim J, Kundu M, Viollet B, Guan K-L. AMPK and mTOR regulate autophagy through direct phosphorylation of Ulk1. Nat Cell Biol. 2011;13(2):132-41.

[121] Shang L, Wang X. AMPK and mTOR coordinate the regulation of Ulk1 and mammalian autophagy initiation. Autophagy. 2011;7(8):924-6.

[122] Pattingre S, Tassa A, Qu X, Garuti R, Xiao HL, Mizushima N, et al. Bcl-2 antiapoptotic proteins inhibit Beclin 1-dependent autophagy. Cell. 2005;122(6):927-39.

[123] Maiuri MC, Le Toumelin G, Criollo A, Rain J-C, Gautier F, Juin P, et al. Functional and physical interaction between Bcl-X(L) and a BH3-like domain in Beclin-1. EMBO J. 2007;26(10):2527-39.

[124] Erlich S, Mizrachy L, Segev O, Lindenboim L, Zmira O, Adi-Harel S, et al. Differential interactions between Beclin 1 and Bcl-2 family members. Autophagy. 2007;3(6):561-8.

[125] Saeki K, Yuo A, Okuma E, Yazaki Y, Susin SA, Kroemer G, et al. Bcl-2 down-regulation causes autophagy in a caspase-independent manner in human leukemic HL60 cells. Cell Death Differ. 2000;7(12):1263-9.

[126] Akar U, Chaves-Reyez A, Barria M, Tari A, Sanguino A, Kondo Y, et al. Silencing of Bcl-2 expression by small interfering RNA induces autophagic cell death in MCF-7 breast cancer cells. Autophagy. 2008;4(5):669-79.

[127] Maiuri MC, Tasdemir E, Criollo A, Morselli E, Vicencio JM, Carnuccio R, et al. Control of autophagy by oncogenes and tumor suppressor genes. Cell Death Differ. 2009;16(1):87-93.

[128] Bellot G, Garcia-Medina R, Gounon P, Chiche J, Roux D, Pouysségur J, et al. Hypoxia-induced autophagy is mediated through hypoxia-inducible factor induction of BNIP3 and BNIP3L via their BH3 domains. Mol Cell Biol. 2009;29(10):2570-81.

[129] Yee KS, Wilkinson S, James J, Ryan KM, Vousden KH. PUMA- and Bax-induced autophagy contributes to apoptosis. Cell Death Differ. 2009;16(8):1135-45.

[130] Vousden KH, Prives C. Blinded by the light: the growing complexity of p53. Cell. 2009;137(3):413-31.

[131] Levine AJ, Oren M. The first 30 years of p53: growing ever more complex. Nat Rev Cancer. 2009;9(10):749-58.

[132] Soussi T, Lozano G. p53 mutation heterogeneity in cancer. Biochem Biophys Res Commun. 2005;331(3):834-42.

[133] Chipuk JE, Kuwana T, Bouchier-Hayes L, Droin NM, Newmeyer DD, Schuler M, et al. Direct activation of Bax by p53 mediates mitochondrial membrane permeabilization and apoptosis. Science. 2004;303(5660):1010-4.

[134] Steele AJ, Prentice AG, Hoffbrand AV, Yogashangary BC, Hart SM, Nacheva EP, et al. P53-mediated apoptosis of CLL cells: evidence for a transcription-independent mechanism. Blood. 2008;112(9):3827-34.

[135] Feng Z, Hu W, de Stanchina E, Teresky AK, Jin S, Lowe S, et al. The regulation of AMPK beta1, TSC2, and PTEN expression by p53: stress, cell and tissue specificity, and the role of these gene products in modulating the IGF-1-AKT-mTOR pathways. Cancer Res. 2007;67(7):3043-53.

[136] Budanov A V., Karin M. p53 target genes Sestrin1 and Sestrin2 connect genotoxic stress and mTOR signaling. Cell. 2008;134(3):451-60.

[137] Maiuri MC, Malik SA, Morselli E, Kepp O, Criollo A, Mouchel PL, et al. Stimulation of autophagy by the p53 target gene Sestrin2. Cell Cycle. 2009;8(10):1571-6.

[138] Crighton D, Wilkinson S, O'Prey J, Syed N, Smith P, Harrison PR, et al. DRAM, a p53-induced modulator of autophagy, is critical for apoptosis. Cell. 2006;126(1): 121-34.

[139] Tasdemir E, Maiuri MC, Galluzzi L, Vitale I, Djavaheri-Mergny M, D'Amelio M, et al. Regulation of autophagy by cytoplasmic p53. Nat Cell Biol. 2008;10(6):676-87.

[140] Morselli E, Shen S, Ruckenstuhl C, Bauer MA, Mariño G, Galluzzi L, et al. p53 inhibits autophagy by interacting with the human ortholog of yeast Atg17, RB1CC1/ FIP200. Cell Cycle. 2011;10(16):2763-9.

[141] Skulachev VP. Bioenergetic aspects of apoptosis, necrosis and mitoptosis. Apoptosis. 2006. p. 473-85.

[142] Yuan J, Lipinski M, Degterev A. Diversity in the mechanisms of neuronal cell death. Neuron. 2003;40(2):401-13.

[143] Azad MB, Chen Y, Gibson SB. Regulation of autophagy by reactive oxygen species (ROS): implications for cancer progression and treatment. Antioxid Redox Signal. 2009;11(4):777-90.

[144] Vandenabeele P, Galluzzi L, Vanden Berghe T, Kroemer G. Molecular mechanisms of necroptosis: an ordered cellular explosion. Nat Rev Mol Cell Biol. 2010;11(10):700-14.

[145] Troyano A, Sancho P, Fernández C, de Blas E, Bernardi P, Aller P. The selection between apoptosis and necrosis is differentially regulated in hydrogen peroxide-treated and glutathione-depleted human promonocytic cells. Cell Death Differ. 2003;10(8):889-98.

[146] Holler N, Zaru R, Micheau O, Thome M, Attinger A, Valitutti S, et al. Fas triggers an alternative, caspase-8-independent cell death pathway using the kinase RIP as effector molecule. Nat Immunol. 2000;1(6):489-95.

[147] Chan FKM, Shisler J, Bixby JG, Felices M, Zheng L, Appel M, et al. A role for tumor necrosis factor receptor-2 and receptor-interacting protein in programmed necrosis and antiviral responses. J Biol Chem. 2003;278(51):51613-21.

[148] Vercammen D, Brouckaert G, Denecker G, Van de Craen M, Declercq W, Fiers W, et al. Dual signaling of the Fas receptor: initiation of both apoptotic and necrotic cell death pathways. J Exp Med. 1998;188(5):919-30.

[149] Fulda S. The mechanism of necroptosis in normal and cancer cells. Cancer Biol Ther. 2013;14(11):999-1004.

[150] Devin A, Cook A, Lin Y, Rodriguez Y, Kelliher M, Liu Z. The distinct roles of TRAF2 and RIP in IKK activation by TNF-R1: TRAF2 recruits IKK to TNF-R1 while RIP mediates IKK activation. Immunity. 2000;12(4):419-29.

[151] Bertrand MJM, Milutinovic S, Dickson KM, Ho WC, Boudreault A, Durkin J, et al. cIAP1 and cIAP2 facilitate cancer cell survival by functioning as E3 ligases that promote RIP1 ubiquitination. Mol Cell. 2008;30(6):689-700.

[152] Bianchi K, Meier P. A tangled web of ubiquitin chains: breaking news in TNF-R1 signaling. Mol Cell. 2009;36(5):736-42.

[153] Wright A, Reiley WW, Chang M, Jin W, Lee AJ, Zhang M, et al. Regulation of early wave of germ cell apoptosis and spermatogenesis by deubiquitinating enzyme CYLD. Dev Cell. 2007;13(5):705-16.

[154] Micheau O, Tschopp J. Induction of TNF receptor I-mediated apoptosis via two sequential signaling complexes. Cell. 2003;114(2):181-90.

[155] Christofferson DE, Yuan J. Necroptosis as an alternative form of programmed cell death. Curr Opin Cell Biol. 2010;22(2):263-8.

[156] Cho Y, Challa S, Moquin D, Genga R, Ray TD, Guildford M, et al. Phosphorylation-driven assembly of the RIP1-RIP3 complex regulates programmed necrosis and virus-induced inflammation. Cell. 2009;137(6):1112-23.

[157] Wu J, Huang Z, Ren J, Zhang Z, He P, Li Y, et al. Mlkl knockout mice demonstrate the indispensable role of Mlkl in necroptosis. Cell Res. 2013;23(8):994-1006.

[158] Sun L, Wang H, Wang Z, He S, Chen S, Liao D, et al. Mixed lineage kinase domain-like protein mediates necrosis signaling downstream of RIP3 kinase. Cell. 2012;148(1-2):213-27.

[159] Cai Z, Jitkaew S, Zhao J, Chiang H-C, Choksi S, Liu J, et al. Plasma membrane translocation of trimerized MLKL protein is required for TNF-induced necroptosis. Nat Cell Biol. 2014;16(1):55-65.

[160] Chen X, Li W, Ren J, Huang D, He W-T, Song Y, et al. Translocation of mixed lineage kinase domain-like protein to plasma membrane leads to necrotic cell death. Cell Res. 2014;24(1):105-21.

[161] Wang H, Sun L, Su L, Rizo J, Liu L, Wang LF, et al. Mixed lineage kinase domain-like protein MLKL causes necrotic membrane disruption upon phosphorylation by RIP3. Mol Cell. 2014;54(1):133-46.

[162] Wang Z, Jiang H, Chen S, Du F, Wang X. The mitochondrial phosphatase PGAM5 functions at the convergence point of multiple necrotic death pathways. Cell. 2012;148(1-2):228-43.

[163] Tenev T, Bianchi K, Darding M, Broemer M, Langlais C, Wallberg F, et al. The ripoptosome, a signaling platform that assembles in response to genotoxic stress and loss of IAPs. Mol Cell. 2011;43(3):432-48.

[164] Feoktistova M, Geserick P, Kellert B, Dimitrova DP, Langlais C, Hupe M, et al. CIAPs block ripoptosome formation, a RIP1/caspase-8 containing intracellular cell death complex differentially regulated by cFLIP isoforms. Mol Cell. 2011;43(3):449-63.

[165] Kaczmarek A, Vandenabeele P, Krysko D V. Necroptosis: the release of damage-associated molecular patterns and its physiological relevance. Immunity. 2013;38(2): 209-23.

[166] Irmler M, Thome M, Hahne M, Schneider P, Hofmann K, Steiner V, et al. Inhibition of death receptor signals by cellular FLIP. Nature. 1997;388(6638):190-5.

[167] Geserick P, Hupe M, Moulin M, Wong WWL, Feoktistova M, Kellert B, et al. Cellular IAPs inhibit a cryptic CD95-induced cell death by limiting RIP1 kinase recruitment. J Cell Biol. 2009;187(7):1037-54.

[168] Oberst A, Dillon CP, Weinlich R, McCormick LL, Fitzgerald P, Pop C, et al. Catalytic activity of the caspase-8-FLIP(L) complex inhibits RIPK3-dependent necrosis. Nature. 2011;471(7338):363-7.

[169] Srinivasula SM, Ashwell JD. IAPs: what's in a name? Mol Cell. 2008;30(2):123-35.

[170] Wang L, Du F, Wang X. TNF-alpha induces two distinct caspase-8 activation pathways. Cell. 2008;133(4):693-703.

[171] He W, Wang Q, Srinivasan B, Xu J, Padilla MT, Li Z, et al. A JNK-mediated autophagy pathway that triggers c-IAP degradation and necroptosis for anticancer chemotherapy. Oncogene. 2013;33(23):3004-13.

[172] He S, Wang L, Miao L, Wang T, Du F, Zhao L, et al. Receptor interacting protein kinase-3 determines cellular necrotic response to TNF-alpha. Cell. 2009;137(6):1100-11.

[173] Vercammen D, Beyaert R, Denecker G, Goossens V, Van Loo G, Declercq W, et al. Inhibition of caspases increases the sensitivity of L929 cells to necrosis mediated by tumor necrosis factor. J Exp Med. 1998;187(9):1477-85.

[174] Duprez L, Takahashi N, Van Hauwermeiren F, Vandendriessche B, Goossens V, Vanden Berghe T, et al. RIP kinase-dependent necrosis drives lethal systemic inflammatory response syndrome. Immunity. 2011;35(6):908-18.

[175] Bohgaki T, Mozo J, Salmena L, Matysiak-Zablocki E, Bohgaki M, Sanchez O, et al. Caspase-8 inactivation in T cells increases necroptosis and suppresses autoimmunity in Bim-/- mice. J Cell Biol. 2011;195(2):277-91.

[176] Günther C, Martini E, Wittkopf N, Amann K, Weigmann B, Neumann H, et al. Caspase-8 regulates TNF-α-induced epithelial necroptosis and terminal ileitis. Nature. 2011;477(7364):335-9.

[177] Feng S, Yang Y, Mei Y, Ma L, Zhu D e., Hoti N, et al. Cleavage of RIP3 inactivates its caspase-independent apoptosis pathway by removal of kinase domain. Cell Signal. 2007;19(10):2056-67.

[178] O'Donnell MA, Perez-Jimenez E, Oberst A, Ng A, Massoumi R, Xavier R, et al. Caspase 8 inhibits programmed necrosis by processing CYLD. Nat Cell Biol. 2011;13(12): 1437-42.

[179] Degterev A, Hitomi J, Germscheid M, Ch'en IL, Korkina O, Teng X, et al. Identification of RIP1 kinase as a specific cellular target of necrostatins. Nat Chem Biol. 2008;4(5):313-21.

[180] Degterev A, Huang Z, Boyce M, Li Y, Jagtap P, Mizushima N, et al. Chemical inhibitor of nonapoptotic cell death with therapeutic potential for ischemic brain injury. Nat Chem Biol. 2005;1(2):112-9.

[181] Lee TH, Shank J, Cusson N, Kelliher MA. The kinase activity of Rip1 is not required for tumor necrosis factor-alpha-induced IkappaB kinase or p38 MAP kinase activation or for the ubiquitination of Rip1 by Traf2. J Biol Chem. 2004;279(32):33185-91.

[182] Van Herreweghe F, Festjens N, Declercq W, Vandenabeele P. Tumor necrosis factor-mediated cell death: to break or to burst, that's the question. Cell Mol Life Sci. 2010;67(10):1567-79.

[183] Los M, Maddika S, Erb B, Schulze-Osthoff K. Switching Akt: from survival signaling to deadly response. BioEssays. 2009;31(5):492-5.

[184] Vandenabeele P, Declercq W, Van Herreweghe F, Vanden Berghe T. The role of the kinases RIP1 and RIP3 in TNF-induced necrosis. Sci Signal. 2010;3(115):re4.

[185] Los M, Mozoluk M, Ferrari D, Stepczynska A, Stroh C, Renz A, et al. Activation and caspase-mediated inhibition of PARP: a molecular switch between fibroblast necrosis and apoptosis in death receptor signaling. Mol Biol Cell. 2002;13(3):978-88.

[186] Los M, Stroh C, Jänicke RU, Engels IH, Schulze-Osthoff K. Caspases: more than just killers? Trends Immunol. 2001;22(1):31-4.

[187] Denecker G, Vercammen D, Steemans M, Vanden Berghe T, Brouckaert G, Van Loo G, et al. Death receptor-induced apoptotic and necrotic cell death: differential role of caspases and mitochondria. Cell Death Differ. 2001;8(8):829-40.

[188] Debatin K-M, Krammer PH. Death receptors in chemotherapy and cancer. Oncogene. 2004;23(16):2950-66.

[189] Herbst RS, Bajorin DF, Bleiberg H, Blum D, Hao D, Johnson BE, et al. Clinical Cancer Advances 2005: major research advances in cancer treatment, prevention, and screening—a report from the American Society of Clinical Oncology. J Clin Oncol. 2006;24(1):190-205.

[190] Tolcher AW, Mita M, Meropol NJ, von Mehren M, Patnaik A, Padavic K, et al. Phase I pharmacokinetic and biologic correlative study of mapatumumab, a fully human monoclonal antibody with agonist activity to tumor necrosis factor-related apoptosis-inducing ligand receptor-1. J Clin Oncol. 2007;25(11):1390-5.

[191] Ichikawa K, Liu W, Zhao L, Wang Z, Liu D, Ohtsuka T, et al. Tumoricidal activity of a novel anti-human DR5 monoclonal antibody without hepatocyte cytotoxicity. Nat Med. 2001;7(8):954-60.

[192] Meng RD, El-Deiry WS. p53-independent upregulation of KILLER/DR5 TRAIL receptor expression by glucocorticoids and interferon-gamma. Exp Cell Res. 2001;262(2):154-69.

[193] Nebbioso A, Clarke N, Voltz E, Germain E, Ambrosino C, Bontempo P, et al. Tumor-selective action of HDAC inhibitors involves TRAIL induction in acute myeloid leukemia cells. Nat Med. 2005;11(1):77-84.

[194] Inoue H, Shiraki K, Ohmori S, Sakai T, Deguchi M, Yamanaka T, et al. Histone deacetylase inhibitors sensitize human colonic adenocarcinoma cell lines to TNF-related apoptosis inducing ligand-mediated apoptosis. Int J Mol Med. 2002;9(5):521-5.

[195] Shetty S, Graham BA, Brown JG, Hu X, Vegh-Yarema N, Harding G, et al. Transcription factor NF-kappaB differentially regulates death receptor 5 expression involving histone deacetylase 1. Mol Cell Biol. 2005;25(13):5404-16.

[196] Lacour S, Micheau O, Hammann A, Drouineaud V, Tschopp J, Solary E, et al. Chemotherapy enhances TNF-related apoptosis-inducing ligand DISC assembly in HT29 human colon cancer cells. Oncogene. 2003;22(12):1807-16.

[197] Ehrhardt H, Fulda S, Schmid I, Hiscott J, Debatin K-M, Jeremias I. TRAIL induced survival and proliferation in cancer cells resistant towards TRAIL-induced apoptosis mediated by NF-kappaB. Oncogene. 2003;22(25):3842-52.

[198] Kim JY, Lee JY, Kim DG, Koo GB, Yu JW, Kim YS. TRADD is critical for resistance to TRAIL-induced cell death through NF-KB activation. FEBS Lett. 2011;585(14): 2144-50.

[199] Green DR, Evan GI. A matter of life and death. Cancer Cell. 2002. p. 19-30.

[200] Gjertsen BT, Bredholt T, Anensen N, Vintermyr OK. Bcl-2 antisense in the treatment of human malignancies: a delusion in targeted therapy. Curr Pharm Biotechnol. 2007;8(6):373-81.

[201] Cotter FE. Unraveling biologic therapy for Bcl-2-expressing malignancies. Semin Oncol. 2004;31(6 Suppl 16):18-21.

[202] Lopes De Menezes DE, Hudon N, McIntosh N, Mayer LD. Molecular and pharmacokinetic properties associated with the therapeutics of Bcl-2 antisense oligonucleotide G3139 combined with free and liposomal doxorubicin. Clin Cancer Res. 2000;6(7): 2891-902.

[203] Bedikian AY, Millward M, Pehamberger H, Conry R, Gore M, Trefzer U, et al. Bcl-2 antisense (oblimersen sodium) plus dacarbazine in patients with advanced melanoma: the oblimersen melanoma study group. J Clin Oncol. 2006;24(29):4738-45.

[204] Milella M, Trisciuoglio D, Bruno T, Ciuffreda L, Mottolese M, Cianciulli A, et al. Trastuzumab down-regulates Bcl-2 expression and potentiates apoptosis induction by Bcl-2/Bcl-XL bispecific antisense oligonucleotides in HER-2 gene-amplified breast cancer cells. Clin Cancer Res. 2004;10(22):7747-56.

[205] Yamanaka K, Rocchi P, Miyake H, Fazli L, Vessella B, Zangemeister-Wittke U, et al. A novel antisense oligonucleotide inhibiting several antiapoptotic Bcl-2 family members induces apoptosis and enhances chemosensitivity in androgen-independent human prostate cancer PC3 cells. Mol Cancer Ther. 2005;4(11):1689-98.

[206] Yamanaka K, Rocchi P, Miyake H, Fazli L, So A, Zangemeister-Wittke U, et al. Induction of apoptosis and enhancement of chemosensitivity in human prostate cancer LNCaP cells using bispecific antisense oligonucleotide targeting Bcl-2 and Bcl-xL genes. BJU Int. 2006;97(6):1300-8.

[207] Oltersdorf T, Elmore SW, Shoemaker AR, Armstrong RC, Augeri DJ, Belli BA, et al. An inhibitor of Bcl-2 family proteins induces regression of solid tumours. Nature. 2005;435(7042):677-81.

[208] Tse C, Shoemaker AR, Adickes J, Anderson MG, Chen J, Jin S, et al. ABT-263: a potent and orally bioavailable Bcl-2 family inhibitor. Cancer Res. 2008;68(9):3421-8.

[209] Wesarg E, Hoffarth S, Wiewrodt R, Kröll M, Biesterfeld S, Huber C, et al. Targeting BCL-2 family proteins to overcome drug resistance in non-small cell lung cancer. Int J Cancer. 2007;121(11):2387-94.

[210] Cragg MS, Harris C, Strasser A, Scott CL. Unleashing the power of inhibitors of oncogenic kinases through BH3 mimetics. Nat Rev Cancer. 2009;9(5):321-6.

[211] Vogler M. Targeting BCL2-proteins for the treatment of solid tumours. Adv Med. 2014(1):1-10.

[212] Roberts AW, Seymour JF, Brown JR, Wierda WG, Kipps TJ, Khaw SL, et al. Substantial susceptibility of chronic lymphocytic leukemia to BCL2 inhibition: results of a phase I study of navitoclax in patients with relapsed or refractory disease. J Clin Oncol. 2012;30(5):488-96.

[213] Tan N, Malek M, Zha J, Yue P, Kassees R, Berry L, et al. Navitoclax enhances the efficacy of taxanes in non-small cell lung cancer models. Clin Cancer Res. 2011;17(6): 1394-404.

[214] Souers AJ, Leverson JD, Boghaert ER, Ackler SL, Catron ND, Chen J, et al. ABT-199, a potent and selective BCL-2 inhibitor, achieves antitumor activity while sparing platelets. Nat Med. 2013;19(2):202-8.

[215] Vandenberg CJ, Cory S. ABT-199, a new Bcl-2-specific BH3 mimetic, has in vivo efficacy against aggressive Myc-driven mouse lymphomas without provoking thrombocytopenia. Blood. 2013;121(12):2285-8.

[216] Touzeau C, Dousset C, Le Gouill S, Sampath D, Leverson JD, Souers a J, et al. The Bcl-2 specific BH3 mimetic ABT-199: a promising targeted therapy for t(11;14) multiple myeloma. Leukemia. 2014;28(1):210-2.

[217] Tam CCS, Seymour JF, Bell A, Westerman DA, Juneja S, Huang DCS, et al. Selective Bcl-2 inhibition with ABT-199 is highly active against chronic lymphocytic leukemia (CLL) irrespective of TP53 mutation or dysfunction. Blood. 2013;122(21):1304.

[218] Vaillant F, Merino D, Lee L, Breslin K, Pal B, Ritchie ME, et al. Targeting BCL-2 with the BH3 mimetic ABT-199 in estrogen receptor-positive breast cancer. Cancer Cell. 2013;24(1):120-9.

[219] Qu Y, Xia P, Zhang S, Pan S, Zhao J. Silencing XIAP suppresses osteosarcoma cell growth, and enhances the sensitivity of osteosarcoma cells to doxorubicin and cisplatin. Oncol Rep. 2014;33(3):1177-84.

[220] Shrikhande S V., Kleeff J, Kayed H, Keleg S, Reiser C, Giese T, et al. Silencing of X-linked inhibitor of apoptosis (XIAP) decreases gemcitabine resistance of pancreatic cancer cells. Anticancer Res. 2006;26(5A):3265-73.

[221] Zhang Y, Wang Y, Gao W, Zhang R, Han X, Jia M, et al. Transfer of siRNA against XIAP induces apoptosis and reduces tumor cells growth potential in human breast cancer in vitro and in vivo. Breast Cancer Res Treat. 2006;96(3):267-77.

[222] Hu Y, Cherton-Horvat G, Dragowska V, Baird S, Korneluk RG, Durkin JP, et al. Antisense oligonucleotides targeting XIAP induce apoptosis and enhance chemotherapeutic activity against human lung cancer cells in vitro and in vivo. Clin Cancer Res. 2003;9(7):2826-36.

[223] Yamaguchi Y, Shiraki K, Fuke H, Inoue T, Miyashita K, Yamanaka Y, et al. Targeting of X-linked inhibitor of apoptosis protein or survivin by short interfering RNAs sensitize hepatoma cells to TNF-related apoptosis-inducing ligand- and chemotherapeutic agent-induced cell death. Oncol Rep. 2005;14(5):1311-6.

[224] Day TW, Najafi F, Wu CH, Safa AR. Cellular FLICE-like inhibitory protein (c-FLIP): a novel target for Taxol-induced apoptosis. Biochem Pharmacol. 2006;71(28):1551-61.

[225] Day TW, Safa AR. RNA interference in cancer: targeting the anti-apoptotic protein c-FLIP for drug discovery. Mini Rev Med Chem. 2009;9(6):741-8.

[226] Ryan BM, O'Donovan N, Duffy MJ. Survivin: a new target for anti-cancer therapy. Cancer Treat Rev. 2009;35(7):553-62.

[227] Bangert A, Cristofanon S, Eckhardt I, Abhari BA, Kolodziej S, Häcker S, et al. Histone deacetylase inhibitors sensitize glioblastoma cells to TRAIL-induced apoptosis by c-myc-mediated downregulation of cFLIP. Oncogene. 2012;31(44):4677-88.

[228] Fulda S. Inhibitor of apoptosis proteins in hematological malignancies. Leukemia. 2009;23(3):467-76.

[229] Fulda S. Targeting inhibitor of apoptosis proteins (IAPs) for cancer therapy. Anti-cancer Agents Med Chem. 2008;8(5):533-9.

[230] Loeder S, Schirmer M, Schoeneberger H, Cristofanon S, Leibacher J, Vanlangenakker N, et al. RIP1 is required for IAP inhibitor-mediated sensitization of childhood acute leukemia cells to chemotherapy-induced apoptosis. Leukemia. 2012;26(5):1742-1742.

[231] Stadel D, Cristofanon S, Abhari BA, Deshayes K, Zobel K, Vucic D, et al. Requirement of nuclear factor κB for Smac mimetic-mediated sensitization of pancreatic carcinoma cells for gemcitabine-induced apoptosis. Neoplasia. 2011;13(12):1162-70.

[232] Carter BZ, Gronda M, Wang Z, Welsh K, Pinilla C, Andreeff M, et al. Small-molecule XIAP inhibitors derepress downstream effector caspases and induce apoptosis of acute myeloid leukemia cells. Blood. 2005;105(10):4043-50.

[233] Probst BL, Liu L, Ramesh V, Li L, Sun H, Minna JD, et al. Smac mimetics increase cancer cell response to chemotherapeutics in a TNF-α-dependent manner. Cell Death Differ. 2010;17(10):1645-54.

[234] Wagner L, Marschall V, Karl S, Cristofanon S, Zobel K, Deshayes K, et al. Smac mimetic sensitizes glioblastoma cells to Temozolomide-induced apoptosis in a RIP1- and NF-κB-dependent manner. Oncogene. 2013;32(8):988-97.

[235] Crawford N, Stasik I, Holohan C, Majkut J, McGrath M, Johnston PG, et al. SAHA overcomes FLIP-mediated inhibition of SMAC mimetic-induced apoptosis in mesothelioma. Cell Death Dis. 2013;4:e733.

[236] Fulda S. Molecular pathways: targeting inhibitor of apoptosis proteins in cancer-from molecular mechanism to therapeutic application. Clin Cancer Res. 2014;20(2): 289-95.

[237] Chauhan D, Neri P, Velankar M, Podar K, Hideshima T, Fulciniti M, et al. Targeting mitochondrial factor Smac/DIABLO as therapy for multiple myeloma (MM). Blood. 2007;109(3):1220-7.

[238] Weisberg E, Kung AL, Wright RD, Moreno D, Catley L, Ray A, et al. Potentiation of antileukemic therapies by Smac mimetic, LBW242: effects on mutant FLT3-expressing cells. Mol Cancer Ther. 2007;6(7):1951-61.

[239] Ong CC, Jubb AM, Haverty PM, Zhou W, Tran V, Truong T, et al. Targeting p21-activated kinase 1 (PAK1) to induce apoptosis of tumor cells. Proc Natl Acad Sci U S A. 2011;108(17):7177-82.

[240] Mehta RSMD, Barlow WEPD, Albain KSMD, Vandenberg TAMD, Dakhil SRMD, Tirumali NRMD, et al. Combination anastrozole and fulvestrant in metastatic breast cancer. N Engl J Med. 2012;367(5):435-44.

[241] Woodcock J, Griffin JP, Behrman RE. Development of novel combination therapies. N Engl J Med. 2011;364(11):985-7.

[242] Fulda S, Vucic D. Targeting IAP proteins for therapeutic intervention in cancer. Nat Rev Drug Discov. 2012;11(2):331-331.

[243] Shingu T, Fujiwara K, Bögler O, Akiyama Y, Moritake K, Shinojima N, et al. Stage-specific effect of inhibition of autophagy on chemotherapy-induced cytotoxicity. Autophagy. 2009;5(4):537-9.

[244] Walker T, Mitchell C, Park MA, Yacoub A, Graf M, Rahmani M, et al. Sorafenib and vorinostat kill colon cancer cells by CD95-dependent and -independent mechanisms. Mol Pharmacol. 2009;76(2):342-55.

[245] Vazquez-Martin A, Oliveras-Ferraros C, Menendez JA. Autophagy facilitates the development of breast cancer resistance to the anti-HER2 monoclonal antibody trastuzumab. PLoS One. 2009;4(7):e6251.

[246] Carew JS, Nawrocki ST, Kahue CN, Zhang H, Yang C, Chung L, et al. Targeting autophagy augments the anticancer activity of the histone deacetylase inhibitor SAHA-to overcome Bcr-Abl-mediated drug resistance. Blood. 2007;110(1):313-22.

[247] Qadir MA, Kwok B, Dragowska WH, To KH, Le D, Bally MB, et al. Macroautophagy inhibition sensitizes tamoxifen-resistant breast cancer cells and enhances mitochondrial depolarization. Breast Cancer Res Treat. 2008;112(3):389-403.

[248] Katayama M, Kawaguchi T, Berger MS, Pieper RO. DNA damaging agent-induced autophagy produces a cytoprotective adenosine triphosphate surge in malignant glioma cells. Cell Death Differ. 2007;14(3):548-58.

[249] Liu D, Yang Y, Liu Q, Wang J. Inhibition of autophagy by 3-MA potentiates cisplatin-induced apoptosis in esophageal squamous cell carcinoma cells. Med Oncol. 2011;28(1):105-11.

[250] Claerhout S, Verschooten L, Van Kelst S, De Vos R, Proby C, Agostinis P, et al. Concomitant inhibition of AKT and autophagy is required for efficient cisplatin-induced apoptosis of metastatic skin carcinoma. Int J Cancer. 2010;127(12):2790-803.

[251] Abedin MJ, Wang D, McDonnell MA, Lehmann U, Kelekar A. Autophagy delays apoptotic death in breast cancer cells following DNA damage. Cell Death Differ. 2007;14(3):500-10.

[252] Roy S, Debnath J. Autophagy and tumorigenesis. Semin Immunopathol. 2010;32(4): 383-96.

[253] Zhou S, Zhao L, Kuang M, Zhang B, Liang Z, Yi T, et al. Autophagy in tumorigenesis and cancer therapy: Dr. Jekyll or Mr. Hyde? Cancer Lett. 2012;323(2):115-27.

[254] Amaravadi RK, Yu D, Lum JJ, Bui T, Christophorou MA, Evan GI, et al. Autophagy inhibition enhances therapy-induced apoptosis in a Myc-induced model of lymphoma. J Clin Invest. 2007;117(2):326-36.

[255] Maclean KH, Dorsey FC, Cleveland JL, Kastan MB. Targeting lysosomal degradation induces p53-dependent cell death and prevents cancer in mouse models of lymphomagenesis. J Clin Invest. 2008;118(1):79-88.

[256] Bellodi C, Lidonnici MR, Hamilton A, Helgason GV, Soliera AR, Ronchetti M, et al. Targeting autophagy potentiates tyrosine kinase inhibitor-induced cell death in Philadelphia chromosome-positive cells, including primary CML stem cells. J Clin Invest. 2009;119(5):1109-23.

[257] Carew JS, Medina EC, Esquivel JA, Mahalingam D, Swords R, Kelly K, et al. Autophagy inhibition enhances vorinostat-induced apoptosis via ubiquitinated protein accumulation. J Cell Mol Med. 2010;14(10):2448-59.

[258] Wu Z, Chang P-C, Yang JC, Chu C-Y, Wang L-Y, Chen N-T, et al. Autophagy blockade sensitizes prostate cancer cells towards Src family kinase inhibitors. Genes Cancer. 2010;1(1):40-9.

[259] Levy JMM, Thorburn A. Targeting autophagy during cancer therapy to improve clinical outcomes. Pharmacol Ther. 2011;131(1):130-41.

[260] Boya P, González-Polo R-A, Casares N, Perfettini J-L, Dessen P, Larochette N, et al. Inhibition of macroautophagy triggers apoptosis. Mol Cell Biol. 2005;25(3):1025-40.

[261] Kanematsu S, Uehara N, Miki H, Yoshizawa K, Kawanaka A, Yuri T, et al. Autophagy inhibition enhances sulforaphane-induced apoptosis in human breast cancer cells. Anticancer Res. 2010;30(9):3381-90.

[262] Hsu KF, Wu CL, Huang SC, Wu CM, Hsiao JR, Yo YT, et al. Cathepsin L mediates resveratrol-induced autophagy and apoptotic cell death in cervical cancer cells. Autophagy. 2009;5(4):451-60.

[263] Li J, Hou N, Faried A, Tsutsumi S, Kuwano H. Inhibition of autophagy augments 5-fluorouracil chemotherapy in human colon cancer in vitro and in vivo model. Eur J Cancer. 2010;46(10):1900-9.

[264] Shimgu T, Fujiwara K, Bogler O, Akiyama Y, Meritake K, Shinojima N, et al. Inhibition of autophagy at a late stage enhances imatinib-induced cytotoxicity in human malignant glioma cells. Int J Cancer. 2009;124(5):1060-71.

[265] Maiuri MC, Zalckvar E, Kimchi A, Kroemer G. Self-eating and self-killing: crosstalk between autophagy and apoptosis. Nat Rev Mol Cell Biol. 2007;8(9):741-52.

[266] Bursch W, Ellinger A, Kienzl H, Török L, Pandey S, Sikorska M, et al. Active cell death induced by the anti-estrogens tamoxifen and ICI 164 384 in human mammary carcinoma cells (MCF-7) in culture: the role of autophagy. Carcinogenesis. 1996;17(8): 1595-607.

[267] Ciuffreda L, Di Sanza C, Incani UC, Milella M. The mTOR pathway: a new target in cancer therapy. Curr Cancer Drug Targets. 2010;10(5):484-95.

[268] Crazzolara R, Bradstock KF, Bendall LJ. RAD001 (everolimus) induces autophagy in acute lymphoblastic leukemia. Autophagy. 2009;5(5):727-8.

[269] Yao JC, Lombard-Bohas C, Baudin E, Kvols LK, Rougier P, Ruszniewski P, et al. Daily oral everolimus activity in patients with metastatic pancreatic neuroendocrine tumors after failure of cytotoxic chemotherapy: a phase II trial. J Clin Oncol. 2010;28(1): 69-76.

[270] Dancey J. mTOR signaling and drug development in cancer. Nat Rev Clin Oncol. 2010;7(4):209-19.

[271] O'Reilly T, McSheehy PMJ, Wartmann M, Lassota P, Brandt R, Lane HA. Evaluation of the mTOR inhibitor, everolimus, in combination with cytotoxic antitumor agents using human tumor models in vitro and in vivo. Anticancer Drugs. 2011;22(1):58-78.

[272] Shaw RJ, Lamia KA, Vasquez D, Koo S-H, Bardeesy N, Depinho RA, et al. The kinase LKB1 mediates glucose homeostasis in liver and therapeutic effects of metformin. Science. 2005;310(5754):1642-6.

[273] Yu K, Shi C, Toral-Barza L, Lucas J, Shor B, Kim JE, et al. Beyond rapalog therapy: preclinical pharmacology and antitumor activity of WYE-125132, an ATP-competitive and specific inhibitor of mTORC1 and mTORC2. Cancer Res. 2010;70(2):621-31.

[274] Feldman ME, Apsel B, Uotila A, Loewith R, Knight ZA, Ruggero D, et al. Active-site inhibitors of mTOR target rapamycin-resistant outputs of mTORC1 and mTORC2. PLoS Biol. 2009;7(2):0371-83.

[275] Thoreen CC, Kang SA, Chang JW, Liu Q, Zhang J, Gao Y, et al. An ATP-competitive mammalian target of rapamycin inhibitor reveals rapamycin-resistant functions of mTORC1. J Biol Chem. 2009;284(12):8023-32.

[276] Janes MR, Limon JJ, So L, Chen J, Lim RJ, Chavez MA, et al. Effective and selective targeting of leukemia cells using a TORC1/2 kinase inhibitor. Nat Med. 2010;16(2): 205-13.

[277] Levine B, Sinha S, Kroemer G. Bcl-2 family members: dual regulators of apoptosis and autophagy. Autophagy. 2008;4(5):600-6.

[278] Voss V, Senft C, Lang V, Ronellenfitsch MW, Steinbach JP, Seifert V, et al. The pan-Bcl-2 inhibitor (-)-gossypol triggers autophagic cell death in malignant glioma. Mol Cancer Res. 2010;8(7):1002-16.

[279] Fulda S, Kögel D. Cell death by autophagy: emerging molecular mechanisms and implications for cancer therapy [Epub ahead of print]. Oncogene. Nature Publishing Group; 2015;

[280] Turcotte S, Chan DA, Sutphin PD, Hay MP, Denny WA, Giaccia AJ. A molecule targeting VHL-deficient renal cell carcinoma that induces autophagy. Cancer Cell. 2008;14(1):90-102.

[281] Zong WX, Ditsworth D, Bauer DE, Wang ZQ, Thompson CB. Alkylating DNA damage stimulates a regulated form of necrotic cell death. Genes Dev. 2004;18(11):1272-82.

[282] Ha HC, Snyder SH. Poly(ADP-ribose) polymerase is a mediator of necrotic cell death by ATP depletion. Proc Natl Acad Sci U S A. 1999;96(24):13978-82.

[283] Jouan-Lanhouet S, Arshad MI, Piquet-Pellorce C, Martin-Chouly C, Le Moigne-Muller G, Van Herreweghe F, et al. TRAIL induces necroptosis involving RIPK1/RIPK3-dependent PARP-1 activation. Cell Death Differ. 2012;19(12):2003-14.

[284] Han W, Li L, Qiu S, Lu Q, Pan Q, Gu Y, et al. Shikonin circumvents cancer drug resistance by induction of a necroptotic death. Mol Cancer Ther. 2007;6(5):1641-9.

[285] Hu X, Han W, Li L. Targeting the weak point of cancer by induction of necroptosis. Autophagy. 2007;3(5):490-2.

[286] Saddoughi SA, Gencer S, Peterson YK, Ward KE, Mukhopadhyay A, Oaks J, et al. Sphingosine analogue drug FTY720 targets I2PP2A/SET and mediates lung tumour suppression via activation of PP2A-RIPK1-dependent necroptosis. EMBO Mol Med. 2013;5(1):105-21.

[287] Deeraksa A, Pan J, Sha Y, Liu X-D, Eissa NT, Lin S-H, et al. Plk1 is upregulated in androgen-insensitive prostate cancer cells and its inhibition leads to necroptosis. Oncogene. 2012;32(24):2973-83.

[288] Pasupuleti N, Leon L, Carraway KL, Gorin F. 5-Benzylglycinyl-amiloride kills proliferating and nonproliferating malignant glioma cells through caspase-independent necroptosis mediated by apoptosis-inducing factor. J Pharmacol Exp Ther. 2013;344(3):600-15.

[289] Dunai ZA, Imre G, Barna G, Korcsmaros T, Petak I, Bauer PI, et al. Staurosporine induces necroptotic cell death under caspase-compromised conditions in U937 cells. PLoS One. 2012;7(7):e41945.

[290] Bonapace L, Bornhauser BC, Schmitz M, Cario G, Ziegler U, Niggli FK, et al. Induction of autophagy-dependent necroptosis is required for childhood acute lymphoblastic leukemia cells to overcome glucocorticoid resistance. J Clin Invest. 2010;120(4):1310-23.

[291] Basit F, Cristofanon S, Fulda S. Obatoclax (GX15-070) triggers necroptosis by promoting the assembly of the necrosome on autophagosomal membranes. Cell Death Differ. 2014;21(7):1183-4.

[292] Laukens B, Jennewein C, Schenk B, Vanlangenakker N, Schier A, Cristofanon S, et al. Smac mimetic bypasses apoptosis resistance in FADD- or caspase-8-deficient cells by priming for tumor necrosis factor α-induced necroptosis. Neoplasia. 2011;13(10):971-9.

[293] Vanlangenakker N, Vanden Berghe T, Bogaert P, Laukens B, Zobel K, Deshayes K, et al. cIAP1 and TAK1 protect cells from TNF-induced necrosis by preventing RIP1/RIP3-dependent reactive oxygen species production. Cell Death Differ. 2011;18(4): 656-65.

[294] Steinhart L, Belz K, Fulda S. Smac mimetic and demethylating agents synergistically trigger cell death in acute myeloid leukemia cells and overcome apoptosis resistance by inducing necroptosis. Cell Death Dis. 2013;4(9):e802.

[295] Pastan I, Hassan R, Fitzgerald DJ KR. Immunotoxin therapy of cancer. Nat Rev Cancer. 2006;6(4):559-65.

[296] Horita H, Frankel AE, Thorburn A. Acute myeloid leukemia-targeted toxin activates both apoptotic and necroptotic death mechanisms. PLoS One. 2008;3(12):e3909.

Chapter 5

Cell Biology of Virus Infection. The Role of Cytoskeletal Dynamics Integrity in the Effectiveness of Dengue Virus Infection

Elizabeth Orozco-García, Andrea Trujillo-Correa and
Juan Carlos Gallego-Gómez

Additional information is available at the end of the chapter

http://dx.doi.org/10.5772/61704

Abstract

The cell biology of viral infections is the focus of this research, in which the role of the cytoskeleton in dengue virus (DENV) replication in cell cultures was evaluated by means of Nocodazole and Cytochalasin D treatments before and after of DENV infection. The potential contribution of cytoskeleton elements with/without the treatment of depolymerizing agents was evidenced and quantified by the subcellular distribution of viral proteins, virions produced, and viral protein quantification. The cytoskeleton is involved in DENV replication because treatments with actin microfilaments and microtubule depolymerizing agents in non-cytotoxic concentrations, affected DENV2 replication in Vero cells and decreased both the viral protein expression and infectious virion production, when compared with non-treated cells. The actin and microtubules are partly involved in DENV2 replication, since the treatment does not completely blocked viral replication, suggesting that these components are necessary but not sufficient alone for DENV2 replication in Vero cells. The structural and functional role of actin and the microtubules in replication are postulated here, opening new perspectives for understanding the architecture of the replicative complex and viral morphogenesis processes, due to the role of the cytoskeleton in the organization, recruitment, and function of the cellular elements necessary for the assembly of viral factories.

Keywords: Cell biology, infection, dengue virus, cytoskeleton, viral replication

1. Introduction

The cell biology of viral infection, as a new scientific approach, considers the cell as the structural and functional unit of viral infection. Within this new research program there are four topics:

1. Virus and the cytoskeleton.
2. Virus and endomembrane systems.
3. Virus and signal cascades.
4. Virus and the nucleus–nuclear envelope [1].

There are a lot of publications already covering these four points of the cell biology of virus infection; however, here we will focus only on the role of the cytoskeleton on DENV infection. Although many years ago these studies could be classified as "virus–host cell interactions" as it was usually considered in many virology journals, since the onset of the new millennium and the advent of converged methodologies/fields of research it is possible to take another view of viral infections. Recent developments in confocal microscopy hardware, image capture and collection with augmented resolution using charge-coupled device (CCD) cameras, digital image processing, improvements in the genetic engineering of green fluorescent protein (GFP) and the chemistry of fluorophores, together with the increased capacity of computational tools for taking thousands of images, have created innovative conditions permitting the understanding of viral infections as a whole [2, 3]. Several researchers consider viruses as tools, or molecular scalpels, which are very useful for dissecting complex molecular and cellular mechanisms [4]. This chapter will consider the cycle of Virus Dengue Serotype 2 (DENV2) as a series of steps in which the virus takes control of the cytoskeleton from host cells [5].

The possibility of real-time visualization of viral infections was reached by means of genetic manipulation of the cells and viruses – cells with fluorescent subcellular compartments and viruses with fluorescent tags [6, 7]; this will be not considered here since the main goal of our approach is to study complete virions within host cells. In this context, molecular studies of isolated viral proteins in a cellular environment are not considered under the cell biology of viral infections discipline, however, those approaches are important if they are following previous research in cell biology of viral infections, because these reductionist studies can offer details regarding the identification of molecular actors in the complex cellular environment of viral infection.

On the other hand, the study of intracellular invaders [5], considering not only viruses but also bacteria, fungi, and protozoan parasites, which aims to find a cellular platform for microbiology research with the eukaryotic host cell as the main actor, is defined as cellular microbiology [8]. Since viruses are obligated parasites, they entirely depend on the cellular machinery to carry out its replication cycle; in this sense for enveloped viruses the intracellular–vesicle trafficking and the cytoskeleton play a decisive role, mainly because both of them offer structural and functional platforms where many RNA viruses are assembled and whose morphogenesis occurs in the cytoplasm of known viral factories [9, 10]. From this viewpoint, DENV will be considered as one experimental system for studying the cell biology of viral infections – its complex interactions could throw light on the search for disease therapies.

Dengue is emerging globally as the most important arboviral disease threatening human populations. It is caused by the *Flavivirus* DENV [11, 12], which is a small and enveloped virus with a single-stranded, positive-sense RNA, which can be translated into a single polyprotein by a host cell and viral proteins [1, 13], and is transmitted to humans by *Aedes* genus mosquitoes [14-17]. There are nearly 2.5 billion people at risk of infection with DENV in tropical and subtropical countries. It is endemic in more than 100 countries and about 100 million cases of dengue fever are estimated annually, with over 500,000 cases being the potentially fatal dengue hemorrhagic fever [12]. Dengue virus is therefore associated with high socio-economic impacts [18, 19].

The availability of vaccines and drugs for RNA viruses has become a much greater challenge than expected, because RNA viruses evolution/mutation rates continue to elude both, vaccine design and drug effectiveness, generating extensive drug-resistant mutants, and therefore over time creating far more virulent strains [20-23]. Therefore, searching for answers inside the cell may offer an alternative method to combat the virus. Recently Villar and collaborators [24], in a big clinical trial implicating several Latin American countries, have shown important protection percentages (greater than 60%), representing an important advance in the fight against the first viral infection in Colombia. In spite of these encouraging clinical responses in patients given the tetravalent vaccine for dengue, it is necessary to wait for several years in order to evaluate the impact of this vaccine has in changing the epidemiological data of the disease. Since such vaccines take time to develop or be assessed, hope could come in the form of the development of antiviral drugs directed to cellular/molecular targets in host cells.

In this way, the cell biology of viral infections [1] has matured considerably in recent years, with cytoskeleton–virus interactions being of particular interest, because it plays both a structural and functional roles in several steps of the viral replication cycle [25, 26], connecting the cell with its environment and participating decisively in the spreading of the virus to neighboring uninfected cells [27]. The virus cannot be transported into the cell by diffusion like small molecules can, therefore, the cytoskeleton and other intracellular structures become a barrier to these [28]. This is why many viruses have shown that viral infection can cause extensive and sophisticated cytoskeletal rearrangements [10, 29-31] and alterations in the endomembrane system, something required for the replication and assembly of new virions [10, 32]. All of this leads to a cytopathic effect, which is widely known and distinctive for viral infection in cellular cultures [32]. This structural and functional intimate interaction between intracellular pathogens and the cell, involves the three elements of the cytoskeleton (actin, microtubules, and intermediate filaments) [30, 33, 34].The involvement of the cytoskeleton in viral infection has been studied for over 30 years [35], and has been found to be involved throughout the life cycle of different viruses. There are many descriptions of its involvement for adenovirus [36], human and equine herpes virus [37], HIV [38], HTLV-1 [27], Rous Sarcoma Virus [39], poliovirus [40], Epstein–Barr Virus [41], human respiratory syncytial virus [42], SV40 [43], Vaccinia virus [44, 45], poliedrosis [46], papiloma [47], and pseudorabies [48]. The replication cycle of DENV has been extensively studied with particular regard to early events such as binding, fusion, uncoating, and intracellular transport of viral proteins [28, 49, 50]. On the other hand, the relationships between the replication of DENV and the cytoskeleton host cell are not fully characterized, although data in some literature, concerning influenza and other viruses, support their existence [13, 51-56].

A structural and functional integration of vesicle trafficking and cytoskeletal/endomembrane systems in viral replication and morphogenesis represents the viral factories redefinition [10, 57, 58]. It has been postulated that trafficking occurs in microtubule motor proteins (entry via dynein and exit through kinesin [59]), something which explains the translocation of proteins between the two routes in the microtubule organizing center (MTOC) [59]. Specifically, with respect to DENV, cytoskeleton and endo/exocytosis have not been studied comprehensively [60], especially where there is dependence between vimentin intermediate filaments and the DENV infection, in which microtubules now were apparently necessary. Intercalating fluorophores were used to label the envelopes of virions associated with cholesterol isolated DENV [50, 61]; tracking was achieved in the early events of cell entry, but this did not shed new light on the endomembrane system and cytoskeleton. It has also been reported that DENV entry by endocytosis, which is pH dependent and mediated by clathrin [62], and the involvement of actin microfilaments and microtubules in mosquito cells DENV had infected was important because the inhibition of these cytoskeleton components decreased infection by 80%. The small Rho GTPases are a protein family which among other cellular functions, govern cytoskeleton reorganization [63]. When isoprenylation of Rho GTPases was blocked, using statins, the incidence of HIV[64] and DENV [65, 66] infection was reduced via alteration of the virus assembly pathway in the rough endoplasmic reticulum.

The interactions between specific viral components and different molecular complexes of the host cell are fundamental in determining the infection rate. By such reasoning, cell biological characterization of the host–pathogen interactions, and knowledge of the possible role of the basic components of the cytoskeleton, microtubules, and actin microfilaments, allow the elucidation of some mechanisms involved in the pathogenesis of DENV. This allows the identification of new potential host targets that could be used to design efficient and rational antiviral therapies.

The applied focus of the cell biology of viral infections is to searching for antiviral therapies based in cellular targets, which have special importance, because the mutation rates of cellular genes are several thousands of times smaller than the antivirals directed to the specific viral proteins. To illustrate this there are several reports of antiviral candidates blocking cellular targets in the host (Src kinase), like the works of Yang and collaborators [67, 68], and some reports from our working group where the HMG-CoA was blocked by lovastatin on assays *in vitro* and *in vivo* [65, 69]. Additionally, we have found out that ubiquitin–proteasome and cytoskeletal elements seem altered after dengue infected cells were treated with curcumin [70]. However, the most relevant finding in this approach is that one protein (IFITM, interferon-induced transmembrane protein) could be altered for blocking different viruses belonging to several families [71].

In order to determine the contribution of the two cytoskeleton elements (actin and microtubules) in DENV replication, a series of experimental assays was completed using Cytochalasin

D and Nocodazole concomitantly using a high preservation/fixation method on the cytoskeleton in order to analyze cellular images taken using fluorescent microscopy in combination with viral titration and quantification of viral proteins. This gave a better understanding of the participation of the cytoskeleton on the functional and structural issues of DENV replication. The data shown here are in agreement with previous reports from several authors confirming distinct roles in the viral infections of DENV [72-76]. The high quality of the subcellular resolution of cytoskeleton elements in this research actually reinforces their role in these structures in DENV replication.

2. Results and discussion

2.1. Preservation/ fixation of the cytoskeleton in infected cells

To understand the cytoskeletons involvement in the DENV replication cycle and virus–host cell interaction, it is necessary to make appropriate preservations with minimal levels of perturbations to the cell morphology. To test which components of the cytoskeleton (microtubules and actin microfilaments) play a role in DENV infection, a method for preserving and fixing the cytoskeleton should be implemented [39, 40], because the conventional preparations for fluorescence microscopy with methanol/acetone, have a very dramatic effect on the cell by flushing (emptying) most of its cellular contents; thus, the information obtained from ultrastructural studies depends on the type of fixation, among other things, and therefore the level of structure conservation [77-79]. In several cell lines it was demonstrated (Fig. 1) that methanol/acetone treatment preserves microtubules but depolymerizes microfilaments and disrupts both mitochondrial and nuclear morphology. PFA-PBS preserves the general cytoarchitecture, but compromises the quality of microstructural detail. PFA-CBS was found to be the best method to fix and preserve cellular microarchitecture (actin, tubulin), without inducing any cellular change.

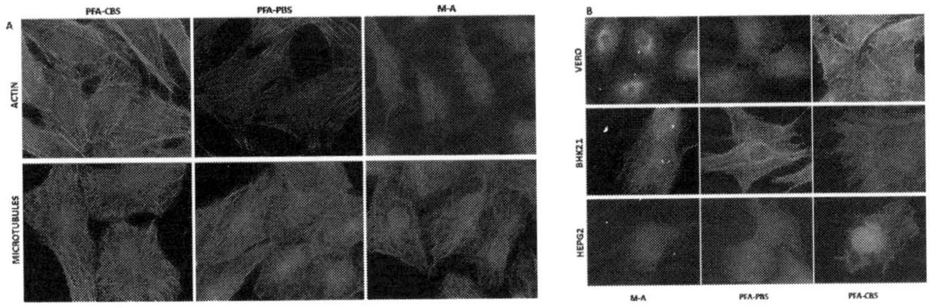

Figure 1. Fluorescence microscopy optimization for visualizing cellular elements. The methanol-acetone (MA) standard protocol does not preserve actin filaments. A comparison between PFA-PBS and PFA-CBS did not demonstrate wide differences in the most notable structures, however, only PFA-CBS method revealed some cellular components which were stabilized satisfactorily. It was found that the optimal concentration for cytoskeletal element visualization was with 3.8% PFA-CBS to fix the cells and antibodies or fluorophores dilutions in CBS of +5% FBS.

2.2. Non-cytotoxic concentrations of cytoskeleton depolymerizing agents

The integrity and functionality of cytoskeleton elements are necessary for their hypothesized role in DENV infections, for which there functionality was evaluated by means of depolymerizing agents. These agents are not entirely harmless to the cells, something which was determined by means dosage ranges having low cytotoxicity (10–20% cytotoxicity) Fig. 2 (a, b). In Fig. 2 (c), the pattern and distribution of cytoskeleton and nucleus components are normal, with no condensation or fragmentation observed. Cellular alterations due to a toxic injury were not evident in these concentrations.

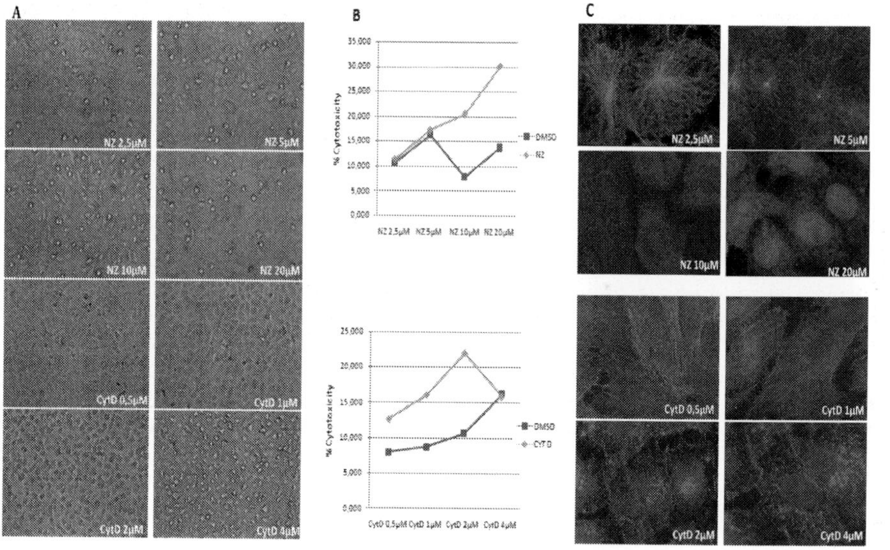

Figure 2. Cytotoxicity assays for cytoskeleton depolymerizing agents. Cytotoxicity assays were performed by cell viability measuring by the method of MTT [3-[4,5 dimetiltiazole-2yl)-2,5 diphenyl tetrazolium bromide), to determine NZ and CytD concentrations that were below the toxic level to cells. We were able to observe the NZ (a, c, e) and CytD (b, d, f) effect in Vero cells. It was found that NZ 10 μM and 4 μM CytD had a cytotoxicity of about 20% and 15%, respectively.

2.3. Relationship between DENV2 and the cytoskeleton

In order to determine some relationship between viral particles and cytoskeletal elements in several steps of the replicative cycle of DENV, Vero cells were infected with DENV2, and processed 18 hours post infection (hpi) by fluorescence microscopy. The fluorescence signal coincident between actin, Fig. 3 (a, b), or the microtubules (c–f), and the viral protein, matching the envelope protein with stress fibers (b) for actin, and microtubules (d), suggests some potential structural relationship, which has an involvement of these cytoskeleton elements within some steps of the DENV infectious cycle.

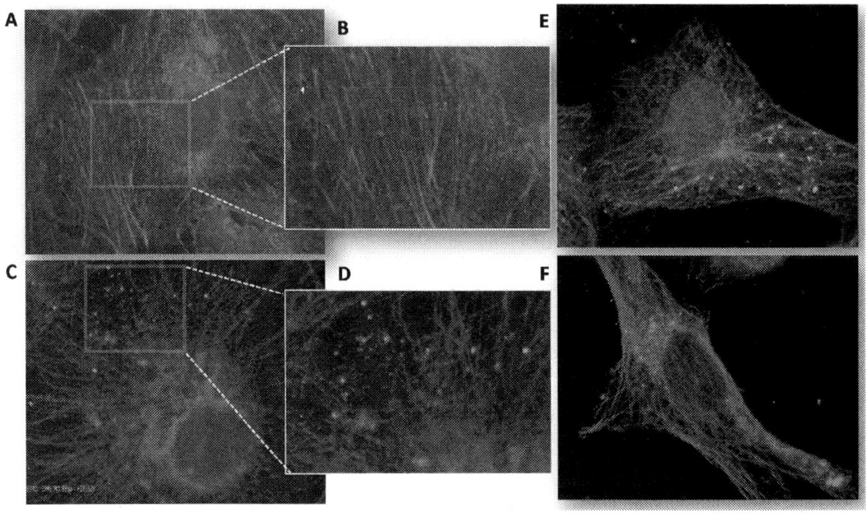

Figure 3. DENV2–cytoskeleton interaction. Vero cell DENV2 infected, viral particles interact with actin (a–b); viral particles interact with microtubules (c–f), were used Phalloidin-Alexa 594 for detect actin or Mab antiβ III Tubulin-Cy3 for microtubules, polyclonal antibody anti-Envelope + antiR IgG Alexa Fluor® 488 for viral envelope protein and Hoechst 33258 for nuclei.

2.4. Subcellular distribution pattern of viral protein altered by depolymerization of microtubules (treatment with nocodazole, NZ)

To investigate the effect of the interrupted cytoskeleton on the viruses replication cycle, Vero cells were treated with NZ before, or after, DENV2 infections. As shown in Fig. 4 (a) and Fig. 5 (a), mock-infected cells without treatment using microtubule depolymerizing agent, NZ, had structures and patterns of observed microtubules traditionally described, i.e., long filaments which are flexible and winding that occupy almost the entire space of the cytosol which emerge from the perinuclear region and are consistent with the MTOC. In contrast, cells that were treated with NZ, Fig. 4 (a), had microtubules which were completely depolymerized, as evidenced by a very fine spotted pattern when observed – something which was not observed the perinuclear region of the MTOC.

Infected cells without treatment with NZ showed a subcellular pattern distribution of diffuse staining of the viral envelope protein, with some small dotted clumps. In order to understand whether microtubules are required for viral entry, treatment was performed 6 hours before infection with NZ (5μM and 10μM), and then with DENV2 at a multiplicity of infection (MOI) of 5 for 1 hour. The inoculum was retired, washings were performed with PBS, and cells were fixed at 24 hpi. The results obtained by fluorescence microscopy indicated that there was a great decrease in the number of infected cells, Fig. 4 (a). In contrast to this, when the cells were infected and post-treated with NZ, significant alterations were produced in this distribution pattern, showed the envelope protein in vesicles as being distributed throughout the cyto-

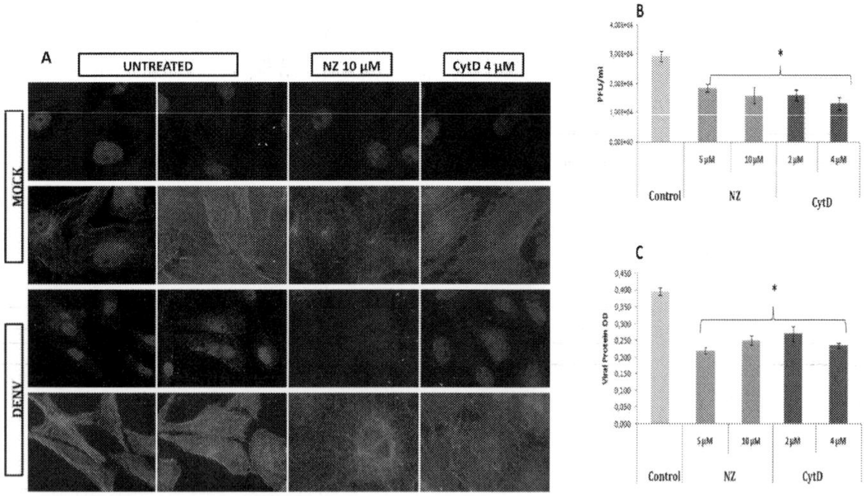

Figure 4. Effects of treatment with NZ and CytD prior to DENV2 infection Vero cells were treated with NZ 5 at 10 µM and CytD 2 at 4 mM, 6 hours before infection to determine if these elements were necessary for virus entry. In both treatments a decrease in infected cells was observed using fluorescence microscopy (a). For NZ and CytD there was an inhibition of about 50% in PFU compared to the control (b), and also a significant decrease of viral protein (c). Were used monoclonal anti-β III Tubulin-Cy3 to detect microtubules, Phalloidin-Alexa 594 for detect actin, polyclonal anti-Envelope + antiR IgG Alexa 488 for viral envelope protein and Hoechst 33258 for nuclei.

plasm, with an apparent subcellular distribution compatible with a Golgi complex Fig. 5 (a). This is because microtubule depolymerization–induced NZ treatment leads to Golgi apparatus disintegration, since the DENV maturation processes used the secretory route in this organelle. In all treatments with NZ, the nucleus was observed with a normal structure.

2.5. Subcellular distribution patterns of viral protein altered by actin depolymerization (treatment with CytD)

To investigate the effect of the interrupted cytoskeleton on the virus replication cycle, Vero cells were treated with cytochalasin D (CytD) before or after infection with DENV2. The mock-infected cells without the depolymerizing agent, CytD, showed a normal pattern with long and parallel fibrils distributed throughout the cytosol, mainly stress fibers, Fig. 4(a) and Fig. 5 (a). This is dramatically reduced after treatment with CytD, resulting in actin depolymerization, and leaving it in a spotted pattern of coarse grains, producing some larger groupings. In the case of actin microfilaments, in DENV2 infected cells, somewhat different findings arose compared with those observed with microtubules. In cells without CytD treatment, remodeling of actin microfilaments was clear, where cells were characterized by *filopodia* formation and many cells were observed with actin tails. These two actin conformations may be related to actin remodeling by Rho GTPases signaling pathway effectors. The envelope viral protein showed two subcellular distribution patterns, one being a diffuse pattern which was very fine

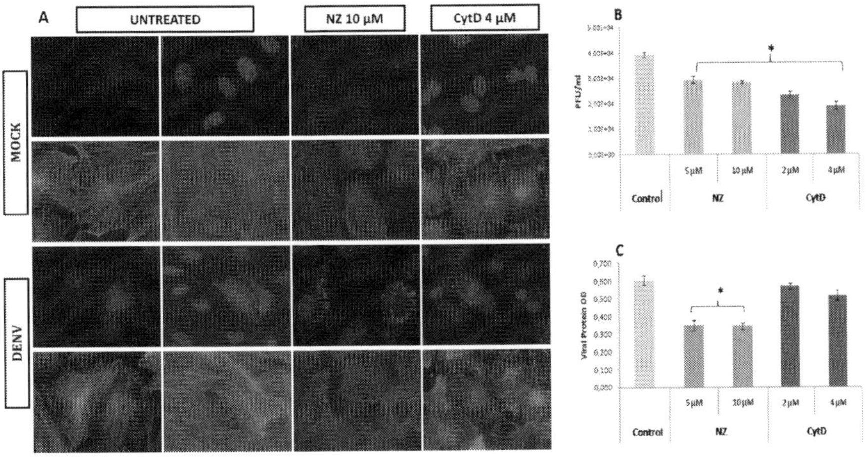

Figure 5. Effects of treatment with NZ and CytD after DENV infection. Vero cells were infected with DENV for 1 hour, the inoculum was removed and washed with PBS, after this the cells were treated with NZ 5–10 μM and CytD 2–4 mM, for 24 hours, in order to determine if these elements were necessary during the replicative cycle of DENV. In both treatments there was a decrease in infected cells when viewed using fluorescence microscopy (a), which was confirmed by the reduction of PFU in the supernatant of the treated cells (b) and viral protein concentration (c). Were used monoclonal anti-β III Tubulin-Cy3 to detect microtubules, phalloidin 594 for actin, polyclonal anti-Envelope + antiR IgG Alexa 488 for viral envelope protein and Hoechst for nuclei

and distributed in the perinuclear region throughout the cytosol, while the other pattern showed a few small clumps of viral protein. According to previous findings in vaccinia virus [45] and DENV [45, 65] the first patters is called "non-assembled viral protein" and the second "assembled viral protein."

In order to understand actin's role in whether microtubules are required for viral entry, the treatment was performed 6 hours before infection with CytD (2μM and 4μM) then with DENV-2 at a MOI of 5 for 1 hour. The inoculum was retired, washings were performed with PBS, and cells were fixed at 24 hpi. To depolymerize actin filaments with CytD in DENV2 infected cells, the subcellular distribution pattern of these cells was altered similarly to uninfected cells. The results obtained by fluorescence microscopy indicated that there was a decrease in the number of infected cells, Fig. 4 (a), with observations using fluorescence microscopy showing a clear accumulation in the perinuclear region that matched the viral factory. In contrast, when cells were infected and post-treated with CytD, significant alterations in this distribution pattern were produced, although the two patterns of distribution of "assembled" viral protein and "non-assembled" changed very little, since they remained in both cells infected with DENV2 and treated with CytD, and in those infected without CytD, Fig. 5 (a). Nevertheless, it is notable that the amount of viral protein "assembled" was markedly reduced in CytD treated cells, suggesting the great role of actin integrity in viral assembly. In all treatments with CytD, the nucleus was observed with a normal structure, Fig. 5 (a).

2.6. The dynamic integrity of microtubules necessary to maintain efficiency of viral replication

To assess if microtubule dynamic integrity has any functional impact on the viral replication process, DENV2-infected cells were pre- or post-treated with 5μM and cell ELISA, after which supernatants were collected in order to quantify viral infectious particle production and intracellular viral protein. NZ pre-treatment, in addition to causing a decrease in the number of infected cells, also produced a decrease in the number of plaque forming units (PFU) quantified in the supernatant with respect to a control group, where the percentage of inhibition od PFUs was 37% for NZ 5 μM and 46% for 10 μM, Fig. 4 (b), and the amount of viral protein in infected cells was reduced to 43% and 37%, respectively, Fig. 4 (c). NZ post-treatment produced a decrease in the number of infectious particles with respect to the control group (infected cells, DMSO), measured by plaque assay (PFU), between 25% and 28% for 5μM and 10uM of this depolymerizing agent, Fig. 5 (b), consistent with a decreased viral protein of 42% approximately for both NZ concentrations, Fig. 5 (c). Together, these results suggest a role is played by microtubules in the assembly, and export, of infectious viral particles. It is probable that, if cellular structures are not properly preserved, not possible observe changes associated with the infection. In this work, we note that altering microtubule dynamics has a direct effect on infection with DENV2, contrary to what was reported by [60].

2.7. Integrity of actin filaments has a greater role in viral protein production than microtubule integrity

To assess if actin dynamic integrity has any functional impact on the viral replication process, DENV2-infected cells were pre- or post-treated with 2μM and 4mM of CytD, after which supernatants were collected for quantifying the viral infectious particle production, and cell ELISA measured intracellular viral protein expressed. In CytD 2–4 μM pre-treated cells, a clear diminution of viral infectious particles was observed, 35–43%, Fig. 4 (b), and viral protein 47–56%, Fig. 4 (c). Interestingly, CytD post-treatment produced a significant decrease in the number of infectious particles production with respect to a control group (infected cells, DMSO), which were measured by plaque assay (PFU), detecting a reduction in viral titer of approximately 37% for CytD 2 μM and 50% for 4 μM, Fig. 5 (b). Actin filament depolymerization produced an insignificant decrease in the amount of viral protein, with non-statistical differences with respect to the control cells, Fig. 5 (c). This suggests that actin may be involved not only in virus entry, but also in any step of virus maturation and, since depolymerized can alter the assembly of viral infectious particles and/or delay the output of these, reflected in a decrease of PFUs and in an intracellular accumulation of viral protein. Thus, although the depolymerization of microtubules and actin filaments reduces the production of infectious viral particles, actin filaments appear to have the most relevant role in this process.

Most recently authors found out that the cytoskeleton is not really necessary for diverse virus replication, specifically for these viruses: Sindbis virus (SINV; family Togaviridae), vesicular stomatitis virus (VSV; family Rhabdoviridae), and Herpes simplex virus (family Herpesviridae) [80]. Although these interesting data suggest a non-conventional role of the cytoskeleton, it must be taken into account that the images are only partially showing the phenomena, because the fine elements of the cytoskeleton (i.e., actin and microtubules), apparently demonstrated some artificial effects due to their preservation/fixation conditions.

There is an important role played by cytoskeleton elements in the assembly and morphogenesis of several viruses, reported in classical works [81] and more recent papers [30, 82, 83], which partially explains the data shown here. The replication complex and the viral factories are cellular structures from viral origins, but these complex structures need macromolecular and cellular constituents. In viral factories membranes and cytoskeleton elements are recruited as part of their constituents, because it is necessary to limit viral activities in this limited space [32]. In this sense, it is logical to attribute the important role of actin and microtubules in the viral assembly process, which in the case of some kind of blocking might produce a reduction in viral particle production.

Although there are some works about the role of the cytoskeleton on DENV infection, in which the participation of filopodia in DENV entry, as well as the function of Rho GTPases in regulation, is confirmed – a process by means of a cross-talk between Rac1 and Cdc42 [74, 76] – the preservation/fixation techniques in those specimens were not the best. Therefore, there are some doubts about the real participation of these elements in the dengue viral infections, because once the cells are significantly damaged the possibility for making the ulterior steps of the infectious cycle are difficult. However, the early stages of viral infections, like those shown by means of filopodia, are undoubtedly because the fine elements are well preserved.

In this vein, there is a collection of data, via images, proteins, and production of virions, demonstrating that the cytoskeleton plays an essential function in the viral replication processes. The viruses can be considered as probes for dissecting cellular signaling, cytoskeleton reorganization, and endocytosis [83, in this sense the research with dengue viruses could be considered similarly – with the actin microfilaments used for the viral assembly process since the disruption of these elements causes envelope viral protein aggregation. Images of "non-assembled" viral proteins have not been shown in others works. This pattern described here is compatible with other viruses [45, 84]. One such piece of work shows that altering actin remodeling by inhibition of a HMG-CoA reductase produces a scattered punctuated fluorescence pattern in viral protein after the treatment of lovastatin, which is compatible with an impaired assembly process. In agreement with these findings, we have shown the DENV2 infected cells treated with lovastatin, produce a clustering of envelope viral protein which also is attributed to some difficulties in the viral assembling process [65].

Consistent with this interpretation, the amounts of viral protein which had accumulated after the inhibition of actin filaments with cytochalasin D treatment was higher in comparison with the microtubules inhibited in the same conditions. If the "non-assembled" protein is reflecting some difficulty in this case [45, 65, 84], the potential role of actin microfilaments is implied in the viral assembly and/or maturation process of DENV. The indirect connection of actin microfilaments, Rho GTPases, with DENV assembly [65] had been used for clinical trials [85], based on the pleiotropic effects of inhibition of HMG-CoA by means of lovastatin [66]. The findings shown here offer new perspectives in the viral replication and morphogenesis of DENV, since at present there is lack of knowledge about the role played by actin in the viral factories and/or in the replication complex. On the other hand, the advantages of the technical approach shown here have been evidenced with other work, in which curcumin is able to alter the DENV2 replication process by blocking several cellular processes such as ubiquitin-proteasome and those of the cytoskeleton [72].

In spite of the diverse roles microtubules and actin microfilaments have in the different viral steps of replication for diverse viruses, more recently it was found that the replicative complex has intermediate filaments of vimentin which have a close, and structural, relationship with the non-structural protein 4A of DENV, forming part of their replication complex [86].

3. Conclusions

There has been an increased interest in recent years in understanding the cellular mechanisms that viruses exploit in the host cell, "Cell Biology of Viral Infection" is a new discipline that seeks to understand these intricate mechanisms [1]. Hence, in order to describe the participation of the cellular elements in DENV2 infection it is necessary to preserve the fine structure of cell morphology as closely as possible by avoiding any generation of artifacts.

In this work we have demonstrated the involvement of the cytoskeleton during DENV replication. It was determined that depolymerized microtubules and actin microfilaments generate disturbances in the DENV2 lifecycle, which causes a reduction in the production of infectious viral particles and in the intracellular expression of the viral envelope protein, as well as an altered subcellular distribution pattern of the viral protein envelope. This effect was more significant in the depolymerization of microtubules that in actin microfilaments. It has been reported that viral infections alter the global host proteomic in response to these, including cytoskeletal proteins [87-89]. However, although a role for the involvement of these proteins in viral infection is hypothesized, a route through which this occurs has yet to be identified.

4. Methods

4.1. Cell lines and virus

Aedes albopictus mosquito C6/36HT cells and epithelial Vero cell line (American Type Culture Collection CCL-81™) were grown in Dulbecco's Eagle's minimum essential medium (DMEM, GIBCO) supplemented with 2% or 10% fetal bovine serum (FBS, GIBCO), with 100 U/ml penicillin/100 μg/ml streptomycin and 0.25 μg/mL Amphotericin B. The cells were grown at 34°C with 5% CO_2 (C6/36HT) and 37°C with 5% CO_2 (Vero). DENV2 (New Guinea strain) was generously donated by María Elena Peñaranda and Eva Harris from the Sustainable Sciences Institute (SSI), San Francisco, California. The viruses were propagated in C6/36HT cells at low multiplicity of infection (MOI: 0,01 PFU/cell) and stored at -80ºC until used. Viral titers were detected by plaque assay, using a Vero cell monolayer culture under 1% methylcellulose overlay medium.

4.2. Antibodies and chemicals

The serum from rabbits against the envelope protein of DENV (polyclonal anti-DENV) was donated by Dr. Eva Harris and Robert Beatty (University of California, Berkeley, CA, USA). The secondary antibody coupled to peroxidase and anti-β-tubulin monoclonal coupled to Cy3 were purchased from Sigma-Aldrich Chemical Co. (St. Louis, MO, USA). Mycotoxin phalloidin conjugated with Alexa Fluor-488 or Alexa Fluor 594 and acetylated tubulin were purchased from Molecular Probes Invitrogen Life Technologies (Carlsbad, CA, USA). CytD, NZ, and 3-(4,5-dimethyl thiazol-2yl)-2,55-diphenyltetrazolium bromide (MTT) were purchased from the Sigma-Aldrich Chemical Co. (St. Louis, MO, USA). FluorSave was purchased from Calbiochem.

4.3. MTT assay

Cell viability was measured by MTT assay [90] to determined NZ and CytD concentrations that were below cytotoxic levels. Cells were seeded, 25,000 cells/well, in 96-well plates and allowed to adhere for 24 hours at 37ºC in 5% CO_2; cells were exposed to MTT (0.5mg/ml) for 4 hours, acid isopropanol (100μl) was then added to solubilize the formazan crystals produced. The culture dishes were incubated for 30 minutes and the absorbance was measured at 570nm in a Benchmark reader (Bio-Rad Laboratories, Hercules, CA). Vero cells were treated with 6 concentrations of these cytoskeleton-depolymerizing agents (NZ: 1.25, 2.5, 5, 10, 20 and 40μM) and (CytD: 0.25, 0.5, 1, 2, 4 and 8μM) for several periods (3, 6, 12, and 24 hours). The NZ and CytD were chosen at two concentrations (5μM and 10μM) and (2μM and 4mM) with a lower cytotoxicity (10% and 15%), but a preserved ability to depolymerize microtubules and inhibit actin polymerization, respectively. These agents were added to cultures 1 hour post-infection (hpi) with DENV.

4.4. Viral infection and treatments

In order to determine the cytoskeletal depolymerizing agent's effects on DENV2 replication, experiments were undertaken as follows. Confluent Vero cell monolayers were grown on 24 (1×10^5) and 96 (2.5×10^4) multi-wells, in DMEM with 2% FBS, at 37°C and 5% CO_2. Cells were infected in the absence of FBS at an MOI of 5 PFU/cell, after a 1-hour adsorption period at 4ºC, the viral inoculum was replaced with maintenance DMEM and incubated at 37ºC. Infected cells were incubated in the presence of NZ, CytD, or DMSO for the indicated times. These cells were maintained for 24 hours after infection and treatment. As a cytoskeleton infection control, we performed the same procedure without DENV under the same conditions. Coverslips for immunofluorescence (IF) were placed in 24 multi-wells. The cell monolayers were fixed and processed to visualize the viruses and cytoskeleton components by confocal fluorescence microscopy. Monolayers and culture supernatants were collected to determine the effect of CytD and NZ in viral protein production by cell ELISA. Additionally, the supernatants also were used to quantify the cytoskeleton blockage effect in infectious viral particle production by means of titration by plaque assay.

4.5. Titration by plaque assays

For quantification of infectious viral particles, the virus was titrated by plaque assays using supernatants of Vero cells infected with DENV2 and treated under different conditions, as described in [91, 92]. Briefly, cells were plated, $5x10^4$ cells/well, on 24-well plates, in DMEM 2% FBS at 37°C in 5% CO_2, and inoculated with serial dilutions of viral collected supernatants from 10^{-1} to 10^{-5} in DMEM without FBS. At 1 hpi the inoculum was removed, washed, had DMEM 2% FBS and finally 1.5% of carboximetilcelulose (SIGMA) added to it. After 8 days post-infection (dpi), the cells were fixed with paraformaldehyde (PFA) at 4% in PBS and revealed with crystal violet. The plaque number was recorded in order to calculate the PFU/ml. Two independent experiments were performed, each in duplicate.

4.6. Cell ELISA

Viral protein was quantified by a spectrophotometric cell ELISA, which is a modification of fluorometric ELISA previously described by [65]. To this immunodetection of viral protein in infected cells, Vero cells ($2.5x10^4$ cells) were submitted to different experimental strategies and were fixed for 30 minutes with 4% paraformaldehyde (PFA), washed three times with PBS, and permeabilized for 30 minutes with 0.1% Triton X-100. Endogenous peroxidase was quenched with 0.3% H_2O_2 in 10% methanol and, non-specific sites were saturated with a blocking buffer (10% FBS in PBS). Then the anti-DENV2 diluted (1:500) blocking buffer was incubated for 1 hour at 37°C. After washing with PBS, this was incubated for 30 minutes with a secondary antibody, anti-mouse IgG conjugated with HRPO. Finally after washing with PBS, chromogenic substrate SIGMA ® FAST OPD (St. Louis, MO, USA) was added to reveal the reaction and its absorbance was read at 405nm in the Benchmark reader (Bio-Rad Laboratories, Hercules, CA, USA). To normalize the data, the total protein concentration in each well was determined by Bradford assay, interpolating absorbance data in a calibration curve with known concentrations of bovine serum albumin. Two independent experiments were performed with 3 replicates for each assay.

4.7. Fluorescence microscopy

In order to use the best preservation and fixation protocol for assays linked to the research, three fixations were assessed: paraformaldehyde 3.7% in phosphate buffered saline (PFA-PBS), paraformaldehyde 3.7% in cytoskeletal buffered sucrose (PFA-CBS), and Methanol-Acetone 1-1 (M-A). Vero cells were cultured in 24 multi-well plates with glass cover slips, at a density of $4.5x10^4$ cells/well in DMEM 2% FBS. After 24 hours monolayers were fixed with different treatments. For the PFA-CBS treatment the medium was discarded, washed with CBS at 37°C, and fixed with PFA-CBS over 30 minutes at 37°C in 5% CO_2. For the PFA-PBS treatment the medium was discarded, washed with PBS, and fixed with PFA-PBS over a period of 30 minutes. For the M-A treatment the medium was discarded, washed with PBS, and fixed with methanol-acetone over a period of 30 minutes at 4°C.

To evaluate the cytoskeleton blocking effects of CytD and NZ in DENV-2 infection, 24 hours after treatment Vero cell monolayers were rinsed once at 37ºC with CBS [10 mM PIPES (pH

6.9) (1,4-piperazinebis-ethane sulfonic acid), 100 mM NaCl, 1.5 mM MgCl2, and 300 mM sucrose. The cells were then permeabilized and simultaneously fixed with PFA-CBS for 30 minutes at 37ºC. To avoid auto-fluorescence the cells were placed in 50mM NH^4Cl for 10 minutes, then were permeabilized with 0.3% Triton X-100 and had their non-specific sites blocked with 5% FBS in CBS. For detection of viral protein and cytoskeleton, cells were incubated with primary monoclonal anti-DENV antibody (1:500) and Alexa 488 conjugated secondary antibody. After that, cells were incubated with anti-phalloidin 594 or anti-β tubulin Cy3 and Hoechst 33258. Following washing with CBS, slides were mounted with mounting medium and examined under a confocal fluorescent microscope (Olympus IX-81 DSU). The micrographs were recorded using an Olympus CCD camera and processed with Image Pro Plus software (from Media Cybernetics).

4.8. Statistical analysis

Data are presented as means ± SD. In all cases, they represent at least three independent determinations. The significance of the results was calculated by Student's t test. In this case, p values < 0.05 were considered significant with respect to controls.

Acknowledgements

This research was supported by COLCIENCIAS grant 111554531592 from the Colombian government. JCGG was the recipient of a Full-Time Professor Program (Exclusive Dedication) for the Medicine Faculty at University of Antioquia for 2014–2015.

Author details

Elizabeth Orozco-García[1], Andrea Trujillo-Correa[2] and Juan Carlos Gallego-Gómez[1*]

*Address all correspondence to: juanc.gallegomez@gmail.com

1 Translational and Molecular Medicine Group, Medellín Medical Research Institute, Faculty of Medicine, Universidad de Antioquia, Medellín, Colombia

2 Institute of Biology, Faculty of Exact and Natural Sciences, Universidad de Antioquia, Medellín, Colombia

The authors declare that they have no competing interests.
AITC contributed to the experimental design, carried out the experiments. EOG worked on analysis interpretation of data, and drafted the manuscript. JCGG conceived the study, participated in its design, and coordinated and finalized the manuscript. All the authors read and approved the final version of the manuscript.

References

[1] Sodeik B, Hannover Medical School DoV, CarlâNeubergstrasse 1, Hannover 30625, Germany, Schramm B, EMBL CBaBP, Meyerhofstrasse 1, Heidelberg 69117, Germany, Suomalainen M, University of Helsinki HI, Department of Virology, PO Box 21, Helsinki FINâ00014, Finland, et al. Meeting Report: EMBO Workshop â˜Cell Biology of Virus Infectionâ™, September 25â"29, 2004, EMBL, Heidelberg, Germany. Traffic. 6(4):351-6.

[2] Radtke K, Dohner K, Sodeik B. Viral interactions with the cytoskeleton: a hitchhiker's guide to the cell. Cell Microbiol. 2006;8(3):387-400.

[3] Stephens DJ, Allan VJ. Light microscopy techniques for live cell imaging. Science. 2003;300(5616):82-6.

[4] Sanderson CM, Smith GL. Cell motility and cell morphology: how some viruses take control. Expert Rev Mol Med. 1999;1999:1-16.

[5] Cellular Microbiology. 2nd ed: ASM Press; 2004. 593 p.

[6] Schudt G, Kolesnikova L, Dolnik O, Sodeik B, Becker S. Live-cell imaging of Marburg virus-infected cells uncovers actin-dependent transport of nucleocapsids over long distances. Proc Natl Acad Sci U S A. 2013;110(35):14402-7.

[7] Lippincott-Schwartz J, Patterson GH. Development and use of fluorescent protein markers in living cells. Science. 2003;300(5616):87-91.

[8] Stephens RS, Sansonetti P, Sibley D. Cellular microbiology - a research agenda and an emerging discipline. Cell Microbiol. 1999;1(1):1-2.

[9] Risco C, Castro IFd, Sanz-Sánchez L, Narayan K, Grandinetti G, Subramaniam S. Three-Dimensional Imaging of Viral Infections. Annual Review of Virology. 2014;1:453-73.

[10] Novoa RR, Calderita G, Arranz R, Fontana J, Granzow H, Risco C. Virus factories: associations of cell organelles for viral replication and morphogenesis. Biol Cell. 2005;97(2):147-72.

[11] Farrar J, Focks D, Gubler D, Barrera R, Guzman MG, Simmons C, et al. Towards a global dengue research agenda. Trop Med Int Health. 12. England2007. p. 695-9.

[12] Guzman MG, Halstead SB, Artsob H, Buchy P, Farrar J, Gubler DJ, et al. Dengue: a continuing global threat. Nature Reviews Microbiology. 2010;8.

[13] Mukhopadhyay S, Kuhn RJ, Rossmann MG. A structural perspective of the flavivirus life cycle. Nat Rev Microbiol. 2005;3(1):13-22.

[14] Gubler DJ, Clark GG. Dengue/dengue hemorrhagic fever: the emergence of a global health problem. Emerg Infect Dis. 1995;1(2):55-7.

[15] Gubler DJ. Dengue and dengue hemorrhagic fever. Clin Microbiol Rev. 1998;11(3): 480-96.

[16] Guzman MG, Kouri G. Dengue diagnosis, advances and challenges. Int J Infect Dis. 2004;8(2):69-80.

[17] Guzman M, Kouri G, Diaz M, Llop A, Vazquez S, Gonzalez D, et al. Dengue, one of the great emerging health challenges of the 21st century. Expert Rev Vaccines. 2004;3(5):511-20.

[18] Halstead SB, Suaya JA, Shepard DS. The burden of dengue infection. Lancet. 2007;369(9571):1410-1.

[19] Shepard DS, Coudeville L, Halasa YA, Zambrano B, Dayan GH. Economic Impact of Dengue Illness in the Americas. Am J Trop Med Hyg. 2011;84(2):200-7.

[20] Mendez JA, Usme-Ciro JA, Domingo C, Rey GJ, Sanchez JA, Tenorio A, et al. Phylogenetic history demonstrates two different lineages of dengue type 1 virus in Colombia. Virol J. 2010;7:226.

[21] Usme-Ciro JA, Mendez JA, Tenorio A, Rey GJ, Domingo C, Gallego-Gomez JC. Simultaneous circulation of genotypes I and III of dengue virus 3 in Colombia. Virol J. 2008;5:101.

[22] Isaguliants MG, Belikov SV, Starodubova ES, Gizatullin RZ, Rollman E, Zuber B, et al. Mutations conferring drug resistance affect eukaryotic expression of HIV type 1 reverse transcriptase. AIDS Res Hum Retroviruses. 2004;20(2):191-201.

[23] Mendez JA, Usme-Ciro JA, Domingo C, Rey GJ, Sanchez JA, Tenorio A, et al. Phylogenetic reconstruction of dengue virus type 2 in Colombia. Virol J. 2012;9:64.

[24] Villar L, Dayan GH, Arredondo-Garcia JL, Rivera DM, Cunha R, Deseda C, et al. Efficacy of a tetravalent dengue vaccine in children in Latin America. N Engl J Med. 2015;372(2):113-23.

[25] Smith GA, Enquist LW. Break ins and break outs: viral interactions with the cytoskeleton of Mammalian cells. Annu Rev Cell Dev Biol. 2002;18:135-61.

[26] Miller S, Krijnse-Locker J. Modification of intracellular membrane structures for virus replication. Nat Rev Microbiol. 2008;6(5):363-74.

[27] Igakura T, Stinchcombe JC, Goon PK, Taylor GP, Weber JN, Griffiths GM, et al. Spread of HTLV-I between lymphocytes by virus-induced polarization of the cytoskeleton. Science. 2003;299(5613):1713-6.

[28] Luby-Phelps K. Cytoarchitecture and physical properties of cytoplasm: volume, viscosity, diffusion, intracellular surface area. Int Rev Cytol. 2000;192:189-221.

[29] Wileman T. Aggresomes and pericentriolar sites of virus assembly: cellular defense or viral design? Annu Rev Microbiol. 2007;61:149-67.

[30] Taylor MP, Koyuncu OO, Enquist LW. Subversion of the actin cytoskeleton during viral infection. Nat Rev Microbiol. 2011;9(6):427-39.

[31] Risco C, Rodriguez JR, Lopez-Iglesias C, Carrascosa JL, Esteban M, Rodriguez D. Endoplasmic reticulum-Golgi intermediate compartment membranes and vimentin filaments participate in vaccinia virus assembly. J Virol. 2002;76(4):1839-55.

[32] Netherton C, Moffat K, Brooks E, Wileman T. A guide to viral inclusions, membrane rearrangements, factories, and viroplasm produced during virus replication. Adv Virus Res. 2007;70:101-82.

[33] Mercer J, Schelhaas M, Helenius A. Virus entry by endocytosis. Annu Rev Biochem. 2010;79:803-33.

[34] Heath CM, Windsor M, Wileman T. Aggresomes Resemble Sites Specialized for Virus Assembly. J Cell Biol. 2001;153(3):449-56.

[35] Bonneau AM, Darveau A, Sonenberg N. Effect of viral infection on host protein synthesis and mRNA association with the cytoplasmic cytoskeletal structure. J Cell Biol. 1985;100(4):1209-18.

[36] Belin MT, Boulanger P. Processing of vimentin occurs during the early stages of adenovirus infection. J Virol. 1987;61(8):2559-66.

[37] Norregard Nielsen L, Forchhammer J, Dabelsteen E, Jepsen A, Stubbe Teglbjaerg C, Norrild B. Herpes simplex virus-induced changes of the keratin type intermediate filament in rat epithelial cells. J Gen Virol. 1987;68 (Pt 3):737-48.

[38] Liu B, Dai R, Tian CJ, Dawson L, Gorelick R, Yu XF. Interaction of the human immunodeficiency virus type 1 nucleocapsid with actin. J Virol. 1999;73(4):2901-8.

[39] Notter MF, Balduzzi PC. Cytoskeletal changes induced by two avian sarcoma viruses: UR2 and Rous sarcoma virus. Virology. 1984;136(1):56-68.

[40] Elliott J, Jones MD, Griffin BE, Krauzewicz N. Regulation of cytoskeletal association by a basic amino acid motif in polyoma virus middle T antigen. Oncogene. 1998;17(14):1797-806.

[41] Friedrich RE, Bartel-Friedrich S, Lobeck H, Niedobitek G, Arps H. Epstein-Barr virus DNA, intermediate filaments and epithelial membrane antigen in nasopharyngeal carcinoma. Anticancer Res. 2000;20(6d):4909-16.

[42] Horvath AR, Kellie S. Regulation of integrin mobility and cytoskeletal association in normal and RSV-transformed chick embryo fibroblasts. J Cell Sci. 1990;97 (Pt 2): 307-15.

[43] Nunbhakdi-Craig V, Craig L, Machleidt T, Sontag E. Simian virus 40 small tumor antigen induces deregulation of the actin cytoskeleton and tight junctions in kidney epithelial cells. J Virol. 2003;77(5):2807-18.

[44] Mallardo M, Schleich S, Krijnse Locker J. Microtubule-dependent Organization of Vaccinia Virus Core-derivd Early mRNAs into Distinct Cytoplasmic Structures. Mol Biol Cell. 2001;12(12):3875-91.

[45] Gallego-Gomez JC, Risco C, Rodriguez D, Cabezas P, Guerra S, Carrascosa JL, et al. Differences in virus-induced cell morphology and in virus maturation between MVA and other strains (WR, Ankara, and NYCBH) of vaccinia virus in infected human cells. J Virol. 2003;77(19):10606-22.

[46] Roncarati R, Knebel-Morsdorf D. Identification of the early actin-rearrangement-inducing factor gene, arif-1, from Autographa californica multicapsid nuclear polyhedrosis virus. J Virol. 1997;71(10):7933-41.

[47] Selinka HC, Giroglou T, Sapp M. Analysis of the infectious entry pathway of human papillomavirus type 33 pseudovirions. Virology. 2002;299(2):279-87.

[48] Van de Walle GR, Favoreel HW, Nauwynck HJ, Van Oostveldt P, Pensaert MB. Antibody-induced internalization of viral glycoproteins in pseudorabies virus-infected monocytes and role of the cytoskeleton: a confocal study. Vet Microbiol. 2002;86(1-2): 51-7.

[49] Mackenzie JM, Jones MK, Westaway EG. Markers for trans-Golgi Membranes and the Intermediate Compartment Localize to Induced Membranes with Distinct Replication Functions in Flavivirus-Infected Cells†. J Virol. 1999;73(11):9555-67.

[50] van der Schaar HM, Rust MJ, Waarts BL, van der Ende-Metselaar H, Kuhn RJ, Wilschut J, et al. Characterization of the early events in dengue virus cell entry by biochemical assays and single-virus tracking. J Virol. 2007;81(21):12019-28.

[51] Berghall H, Wallen C, Hyypia T, Vainionpaa R. Role of cytoskeleton components in measles virus replication. Arch Virol. 2004;149(5):891-901.

[52] Douglas MW, Diefenbach RJ, Homa FL, Miranda-Saksena M, Rixon FJ, Vittone V, et al. Herpes simplex virus type 1 capsid protein VP26 interacts with dynein light chains RP3 and Tctex1 and plays a role in retrograde cellular transport. J Biol Chem. 2004;279(27):28522-30.

[53] Lakadamyali M, Rust MJ, Babcock HP, Zhuang X. Visualizing infection of individual influenza viruses. Proc Natl Acad Sci U S A. 2003;100(16):9280-5.

[54] Leopold PL, Kreitzer G, Miyazawa N, Rempel S, Pfister KK, Rodriguez-Boulan E, et al. Dynein- and microtubule-mediated translocation of adenovirus serotype 5 occurs after endosomal lysis. Hum Gene Ther. 2000;11(1):151-65.

[55] Martinez-Moreno M, Navarro-Lerida I, Roncal F, Albar JP, Alonso C, Gavilanes F, et al. Recognition of novel viral sequences that associate with the dynein light chain LC8 identified through a pepscan technique. FEBS Lett. 2003;544(1-3):262-7.

[56] Ng ML, Hong SS. Flavivirus infection: essential ultrastructural changes and association of Kunjin virus NS3 protein with microtubules. Arch Virol. 1989;106(1-2):103-20.

[57] Ward BM. The Taking of the Cytoskeleton One Two Three: How Viruses Utilize the Cytoskeleton During Egress. Virology. 2011;411(2):244-50.

[58] Netherton CL, Wileman T. Virus factories, double membrane vesicles and viroplasm generated in animal cells. Current Opinion in Virology. 2011;1(5):381-7.

[59] Pierini R, Cottam E, Roberts R, Wileman T. Modulation of membrane traffic between endoplasmic reticulum, ERGIC and Golgi to generate compartments for the replication of bacteria and viruses. Semin Cell Dev Biol. 2009;20(7):828-33.

[60] Chen W, Gao N, Wang JL, Tian YP, Chen ZT, An J. Vimentin is required for dengue virus serotype 2 infection but microtubules are not necessary for this process. Arch Virol. 2008;153(9):1777-81.

[61] van der Schaar HM, Rust MJ, Chen C, van der Ende-Metselaar H, Wilschut J, Zhuang X, et al. Dissecting the cell entry pathway of dengue virus by single-particle tracking in living cells. PLoS Pathog. 2008;4(12):e1000244.

[62] Mosso C, Galvan-Mendoza IJ, Ludert JE, del Angel RM. Endocytic pathway followed by dengue virus to infect the mosquito cell line C6/36 HT. Virology. 2008;378(1): 193-9.

[63] Heasman SJ, Ridley AJ. Mammalian Rho GTPases: new insights into their functions from in vivo studies. Nat Rev Mol Cell Biol. 2008;9(9):690-701.

[64] del Real G, Jimenez-Baranda S, Mira E, Lacalle RA, Lucas P, Gomez-Mouton C, et al. Statins Inhibit HIV-1 Infection by Down-regulating Rho Activity. J Exp Med. 2004;200(4):541-7.

[65] Martinez-Gutierrez M, Castellanos JE, Gallego-Gomez JC. Statins reduce dengue virus production via decreased virion assembly. Intervirology. 2011;54(4):202-16.

[66] Rothwell C, Lebreton A, Young Ng C, Lim JY, Liu W, Vasudevan S, et al. Cholesterol biosynthesis modulation regulates dengue viral replication. Virology. 2009;389(1-2): 8-19.

[67] Yang CC, Tsai MH, Hu HS, Pu SY, Wu RH, Wu SH, et al. Characterization of an efficient dengue virus replicon for development of assays of discovery of small molecules against dengue virus. Antiviral Res. 2013;98(2):228-41.

[68] Chu JJH, Yang PL. c-Src protein kinase inhibitors block assembly and maturation of dengue virus. Proc Natl Acad Sci U S A. 2007;104(9):3520-5.

[69] Martinez-Gutierrez M, Correa-Londono LA, Castellanos JE, Gallego-Gomez JC, Osorio JE. Lovastatin delays infection and increases survival rates in AG129 mice infected with dengue virus serotype 2. PLoS One. 2014;9(2):e87412.

[70] Padilla SL, Rodriguez A, Gonzales MM, Gallego GJC, Castano OJC. Inhibitory effects of curcumin on dengue virus type 2-infected cells in vitro. Arch Virol. 2014;159(3): 573-9.

[71] Brass AL, Huang IC, Benita Y, John SP, Krishnan MN, Feeley EM, et al. The IFITM proteins mediate cellular resistance to influenza A H1N1 virus, West Nile virus, and dengue virus. Cell. 2009;139(7):1243-54.

[72] Padilla-S L, Rodríguez A, Gonzales MM, Gallego-G JC, Castaño-O JC. Inhibitory effects of curcumin on dengue virus type 2-infected cells in vitro. Arch Virol. 2013.

[73] Yang J, Zou L, Hu Z, Chen W, Zhang J, Zhu J, et al. Identification and characterization of a 43 kDa actin protein involved in the DENV-2 binding and infection of ECV304 cells. Microbes Infect. 2013;15(4):310-8.

[74] Zamudio-Meza H, Castillo-Alvarez A, Gonzalez-Bonilla C, Meza I. Cross-talk between Rac1 and Cdc42 GTPases regulates formation of filopodia required for dengue virus type-2 entry into HMEC-1 cells. J Gen Virol. 2009;90(Pt 12):2902-11.

[75] Shrivastava N, National Institute of Virology P, India, Sripada S, National Institute of Virology P, India, Kaur J, National Institute of Virology P, India, et al. Insights into the Internalization and Retrograde Trafficking of Dengue 2 Virus in BHK-21 Cells. PLOS ONE. 2011;6(10).

[76] Wang JL, Zhang JL, Chen W, Xu XF, Gao N, Fan DY, et al. Roles of Small GTPase Rac1 in the Regulation of Actin Cytoskeleton during Dengue Virus Infection. PLoS Negl Trop Dis. 2010;4(8).

[77] Cande WZ, Lazarides E, McIntosh JR. A comparison of the distribution of actin and tubulin in the mammalian mitotic spindle as seen by indirect immunofluorescence. J Cell Biol. 1977;72(3):552-67.

[78] Connolly JA, Kalnins VI, Cleveland DW, Kirschner MW. Immunoflourescent staining of cytoplasmic and spindle microtubules in mouse fibroblasts with antibody to tau protein. Proc Natl Acad Sci U S A. 1977;74(6):2437-40.

[79] Szechyńska-Hebda M, University W, Sciences IoPPPAo, Wędzony M, Sciences IoPPPAo, Dubas E, et al. Visualisation of microtubules and actin filaments in fixed BY-2 suspension cells using an optimised whole mount immunolabelling protocol. Plant Cell Reports. 2006;25(8):758-66.

[80] Matthews JD, Morgan R, Sleigher C, Frey TK. Do viruses require the cytoskeleton? Virol J. 2013;10:121.

[81] Genty N, Bussereau F. Is cytoskeleton involved in vesicular stomatitis virus reproduction? J Virol. 1980;34(3):777-81.

[82] Hyde JL, Gillespie LK, Mackenzie JM. Mouse norovirus 1 utilizes the cytoskeleton network to establish localization of the replication complex proximal to the microtubule organizing center. J Virol. 2012;86(8):4110-22.

[83] Jeffree CE, Brown G, Aitken J, Su-Yin DY, Tan BH, Sugrue RJ. Ultrastructural analysis of the interaction between F-actin and respiratory syncytial virus during virus assembly. Virology. 2007;369(2):309-23.

[84] Ravi LI, Liang L, Wong PS, Brown G, Tan BH, Sugrue RJ. Increased hydroxymethylglutaryl coenzyme A reductase activity during respiratory syncytial virus infection mediates actin dependent inter-cellular virus transmission. Antiviral Res. 2013;100(1):259-68.

[85] Whitehorn J, Van Vinh Chau N, Truong NT, Tai LT, Van Hao N, Hien TT, et al. Lovastatin for adult patients with dengue: protocol for a randomised controlled trial. Trials. 2012;13:203.

[86] Teo CS CJ. Cellular Vimentin Regulates Construction of Dengue V... [J Virol. 2014] - PubMed - NCBI J Virol.2014. Available from: http://www.ncbi.nlm.nih.gov/pubmed/.

[87] Frontiers: Global host proteomic responses to virus infection: Frontiers Media SA; 2014.

[88] Zhang M, Zheng X, Wu Y, Gan M, He A, Li Z, et al. Differential proteomics of Aedes albopictus salivary gland, midgut and C6/36 cell induced by dengue virus infection. Virology. 2013;444(1-2):109-18.

[89] Pando-Robles V, Oses-Prieto JA, Rodriguez-Gandarilla M, Meneses-Romero E, Burlingame AL, Batista CV. Quantitative proteomic analysis of Huh-7 cells infected with Dengue virus by label-free LC-MS. J Proteomics. 2014;111:16-29.

[90] Mosmann T. Rapid colorimetric assay for cellular growth and survival: application to proliferation and cytotoxicity assays. J Immunol Methods. 1983;65(1-2):55-63.

[91] Meiklejohn G, England B, Lennette. Propagation of dengue virus strains in unweaned mice. Am J Trop Med Hyg. 1952;1(1):51-8.

[92] Fujita N, Tamura M, Hotta S. Dengue virus plaque formation on microplate cultures and its application to virus neutralization (38564). Proc Soc Exp Biol Med. 1975;148(2):472-5.

Chapter 6

Autophagy and Lipid Metabolism – A Cellular Platform where Molecular and Metabolic Pathways Converge to Explain Dengue Viral Infection

Elizabeth Orozco-García and Juan Carlos Gallego-Gómez

Additional information is available at the end of the chapter

http://dx.doi.org/10.5772/61305

Abstract

Dengue virus (DENV) is one of the most prevalent human pathogens worldwide. It causes a huge socioeconomic burden with approximately 400 million infections per year, but yet there is no vaccine or antiviral that is currently effective against the disease. DENV is spread by the mosquitoes *Aedes aegypti* and *Aedes albopictus*, and viral replication within the mosquito vector is required for transmission to human host. During its replication cycle, the virus cause significant changes to the host transcriptome profile, especially in the metabolic and trafficking pathways. Recent studies have shown a strong association between autophagy and lipid metabolism modulation.

For many years, biochemistry studies have been forgotten and replaced by the most advanced techniques and theories in molecular biology and their promises for solving the "life code"; however, after many years of strong molecular biology research, it had not found the key of many problems with which we have the elemental biosystems like viruses. Decades of molecular virology investigations did not give more light about several cellular processes that occurred into the host cells when the infections happen. The molecular virologists have cloned many viral genes, manipulating full viral genomes, and engineering chimeric constructs to study many details at the molecular level, but the host cell and the encrypted viruses do not want to reveal their secrets.

Only with the new perspective of complex diseases, a new approach has emerged: An integrative methodology wherein molecular cell biology is converging with the most pure and elegant biochemistry. In this way, more extensive research is necessary for future comparative analyses of the host and vector metabolic/signaling environments required for viral replication.

Keywords: Autophagy, Cellular Platform, Dengue Virus, Molecular and Metabolic Pathways

1. Introduction

1.1. Dengue Virus (DENV): Some clinic and basic issues

Emergent viruses with major impact in human health include several agents of Flavivirus gender, *Flaviviridae* family, the most important of these are DENV (Dengue Virus), YFV (Yellow Fever Virus), JEV (Japanese Encephalitis Virus) and WNV (West Nile Virus) [1]. There are nearly 3.6 billion people at risk of infection with DENV in tropical and subtropical countries [2]. In more than 100 endemic countries with an estimated nearly 390 million of DENV infections per year, approximately 100 million of dengue fever cases are estimated annually with over 2 millions cases of potentially fatal dengue hemorrhagic fever [3, 4]. In most cases, the symptoms of DF that include an acute febrile illness with retro-orbital pain, myalgia, arthralgia are self-limited [5]. However, in a proportion of people, the disease progresses to the severe clinical manifestations classified as dengue shock syndrome (DSS), which are characterized by the plasma leakage leading to hypovolemic shock and/or dengue hemorrhagic fever (DHF), which are characterized by massive bleeding, thrombocytopenia, evidence of plasma leakage such as pleural effusion and a rise of hematocrit, both of which has a high mortality rate [6-8].

DENV is a positive-single strand RNA virus surrounded by an icosahedral nucleocapsid (C) with approximately 10,700 bases, a unique open reading frame that codify to one polyprotein, which is post-translational cleaved by cellular and viral proteases. The 5′ end contains the region encoding the structural proteins in the following order: core protein (protein C), membrane precursor protein (protein M), and envelope protein (protein E). The remainder of it genome encode for seven non-structural proteins, xlink, NS2A, NS2B, NS3, NS4A, NS4A, and NS5 [9].

DENV exists as a four serotypes (DENV-1, DENV-2, DENV-3, and DENV-4). All of them have a same transmission cycle, which include vertebrate hosts (primate and human) and invertebrate vectors, mosquitoes of the following species: *Aedes aegypti, A. albopictus y A. polynesiensis* mosquitoes. An infection with one serotype provides lifelong protective immunity to that serotype. But, there is no cross-protective immunity between serotypes [10, 11]. Inside each one of these serotypes, there are several virus groups named genotypes.

2. Autophagy

This is defined in a general form like a catabolic selective process by means of which cytoplasmic material is transported to lysosomes for their degradation [12]. The autophagy is a remarkably conserved cellular process, from yeast to human, responsible for removing damaged organelles and misfolded proteins, and for maintaining cellular homeostasis under both normal and stress conditions [13-15]. Compartmentalization in eukaryotic systems brought numerous evolutionary advantages, but also great and new challenges with it, such as the selective removal of damaged organelles, controlled organelle number and quality, or the utilization of their components as potential energy source during times of starvation. In this way, autophagy represents an evolutionary answer to these challenges. It enables the recycling of intracellular components and allows cells to survive or death [16] (Figure 1).

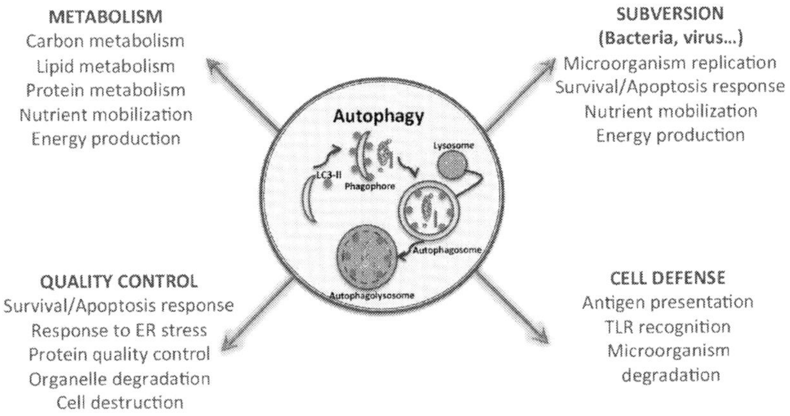

Figure 1. Functions of autophagy

The primary role of autophagy is to protect cells under stressful conditions. Under this viewpoint, both autophagy and the vertebrate immune system play essential roles to maintain cellular homeostasis in the face of external perturbations [17]. Indeed, several studies have revealed the narrow relationship between autophagy and the vertebrate immune system [18]. Besides, the crosstalk has become evident between autophagy and apoptosis [19-22] because the induction of autophagy has often been linked to inhibition of apoptosis [23].

More than 30 genes have been identified as crucial in the autophagy regulation process in yeast, which are known as ATG (autophagy-related genes). Many of these genes have homologs in mammals and are grouped according to expression and participation in the different stages of the autophagic route [12]. The activation of this pathway depends on the kinase mTOR (mammalian Target of Rapamycin) identified as the main negative regulator when the cell is in the presence of growth factors and abundance of nutrients. Under starvation, mTOR activity is inhibited. And consequently, the autophagy is activated allowing the recruitment of complexes inducers of the route [24, 25]. There are three mechanisms identified for autophagic degradation: macroautophagy, microautophagy, and chaperone-mediated autophagy (CMA).

Macroautophagy imply the formation of double membrane vesicles recognized as autophagosomes. It engulfs cytoplasmic components and then are fused to lysosomes, carrying the cytosolic material until the lysosomal lumen where a hydrolases, lipases, and cathepsins degrade it [26, 27]. Therefore, we can divide the pathway into 4 basic steps: initiation, elongation, termination, and fusion. During initiation, the recruitment of protein complexes, such as phosphatidylinositol-3-kinase Class III (PI3K), Vps15, Vps34, and beclin-1, that are critical to the autophagosome formation is given [28]. During the elongation, the assembly occurs. In this stage, the related protein complex ATG5-Atg12-Atg16, the lipid conjugation complex LC3-II- phosphatidylethanolamine (PE), and the respective conjugating enzymes, which act similarly to ubiquitin ligase system, link lipid that allows the growth of the double membrane due to the transformation that undergone LC3-I to LC3-II, which has a PE-binding domain, the main lipid component of autophagosomes. Later, in the termination stage, the double membrane vesicle is closed with the intracytoplasmic content therein, which is possible because the cut that performs Atg-4 enzyme on the binding LC3-II-PE permits the release of the complex into the cytosol preventing the continuation of joining new lipid molecules [28].

Subsequently, the fusion process occurs between autophagosomes and lysosomes, which generate a vesicular structure called autophagolysosome or autolysosome. This process is mediated by the cell membrane fusion proteins, such as integral proteins SNARE (soluble N-Ethylmaleimide-sensitive factor-attachment protein receptor) [29], the Rab family proteins, especially Rab7 and Rab9, that are involved in the transport of the vesicles and fusion with target membranes [30]. And besides, the lysosomal membrane receptor LAMP2 allows the attachment between vesicular membranes and autophagosome contents discharge into the lysosome forming the structure known as autophagolysosome, where protein degradation occurs [31, 32] (Figure 2).

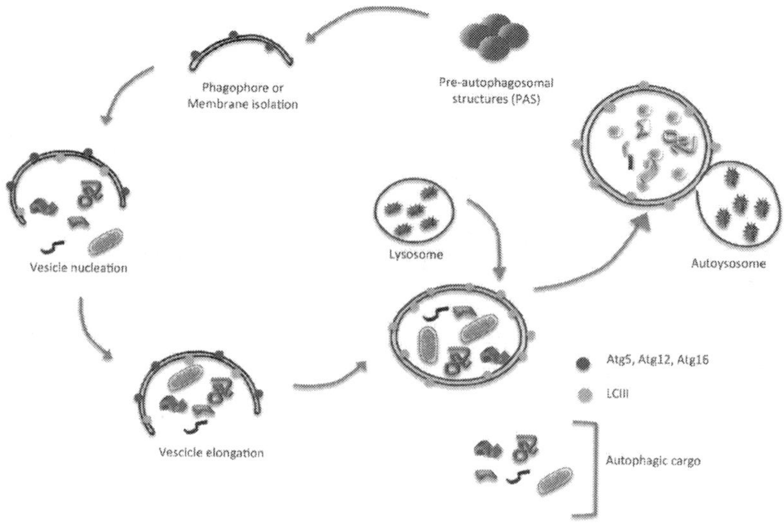

Figure 2. The Process of Macroautophagy. A portion of cytoplasm (including organelles) is enclosed by a phagophore or isolation membrane to form an autophagosome. The outer membrane of the autophagosome subsequently fuses with the lysosome, and the internal material is degraded in the autolysosome.

http://dx.doi.org/10.5772/61305

The second possible mechanism is microautophagy. It involves direct lysosomal membrane invaginations production to generate vesicles or tubules capturing surrounding cytoplasm. Microautophagy studies has mainly been developed in yeast, wherein several genes have been found sharing the macroautophagy and microautophagy pathways, but the components and regulation systems are still not well understood [33]. For the CMA mechanism, the cargo is specifically selected by the unique pentapeptide motif present in the amino acid sequence (KFERQ), which is recognized by the chaperone proteins specially Hsc70, in where the membrane receptor LAMP2 (Lysosome associated membrane protein 2] carrying the load into the lysosome lumen for degradation [34].

In the past years, autophagy has emerged as a critical player in the control of viral infection and immunity [35-39]. On one hand, autophagy can serve as a host defense mechanism for some pathogens by clearing them out of the cells [40-42]. On the other hand, many positive-stranded RNA viruses have been reported to subvert this cellular machinery to favor their own replication and release [23]. This issue will be discussed below.

3. Cellular metabolism in viral infections: Rediscovering the other side of the coin

Metabolism is broadly defined as the sum of biochemical processes in living organisms that either produce or consume energy [43]. In the "Golden Age of Biochemistry" (1920s to 1960s), most of the metabolic network in humans and other organisms, which included routes like glycolysis (Embden, Meyerhof, and Parnas), respiration (Warburg), the tricarboxylic acid (TCA) and urea cycles (Krebs), glycogen catabolism (Cori and Cori), oxidative phosphorylation (Mitchell), and the supremacy of ATP in energy transfer reactions (Lippmann) was defined. This research was awarded with about 15 Nobel Prizes in Physiology, Medicine, or Chemistry. All of them were related to energy balance or core metabolic pathways [43].

Richard W. Hanson wrote "By 1970, the writing was on the wall for metabolism; it was largely considered a "mature area", lacking excitement; molecular biology was the area of the future" [44].

"A sure sign of this was that graduate students in biochemistry almost never selected their thesis research in metabolism. The course in intermediary metabolism that I taught was dropped from the curriculum of our graduate education program; our students were expected to learn all they needed to know about metabolism as undergraduates before they attended graduate school. After all, as a graduate student once said to me, "the great problems in metabolic research have been solved". As long as diseases like diabetes, obesity, and atherosclerosis, remain to be cured, there will be no shortage of interest in metabolism" [44].

In this way, the understanding of diseases in light of alterations in metabolic status was dropped and shifted by the search of an explanation based on the nascent era of molecular biology. However, the ongoing exploration of molecular biology and disease complexity has stimulated a revival of interest in intermediary metabolism [45]. In this view, several works propose a new way to arrive the disease: cell metabolism, because it affects cell signaling and modulate protein trafficking, localization, and enzyme activity [43]. For example, Acetyl-CoA plays a central role in intermediary metabolism (carbohydrate, fatty acid, and amino acid oxidation,) and at the same time have tremendous influence on cell signaling and gene expression [46-48]. Recently, it has been demonstrated that some biomarkers of metabolic syndrome are related with any infection, acute or chronic in patients [49].

To reach a deep and elegant comprehension of the role of metabolism in all levels of the human being, it is better to take the exact quotation of DeBerardinis and Thompson: "...the metabolism pervades every aspect of biology from the single-cell to whole organism level. No cellular functions occur independently of metabolism, and a metabolic perturbation at one node has ripple effects that can extend throughout the network and out into other systems. Thus, metabolic disturbances have an extremely long reach, and this extends to disease phenotypes..." [43].

The Warburg effect is a concept used to link metabolism and cancer wherein a disturbance of cellular metabolic activity is at the root of tumor formation and growth [43]. Thereby, dysregulated cellular metabolism is a key feature of cancer [50-53]. This concept could be adapted perfectly to viral infections, because viruses are biological entities that depend on cell metabolism to replicate and spread. Therefore, it would be expected that the success of viruses inside the cell will be dependent on their ability to subrogate the metabolism and put it in his favor.

It had been shown in this sense that tumor cells display increased metabolic autonomy in comparison to non-transformed cells [51]. In the case of viral infections, this "metabolic autonomy" may be triggered by a viral entity in normal cells. Thinking about it, it is not absurd if we take into account that many genes implicated in several signaling/metabolic pathways have also been reported to be modulated and altered in viral infections [54-57]. How these metabolic pathways are regulated in infected cells, including if they fluctuate according to infection stage or at the cell cycle, remains to be a question. It will be important to determine whether viral infection can regulate all aspects of the metabolic dynamics or if any special metabolic pathway implicated in their replication or pathogenicity exists.

Many pathogens have developed sophisticated molecular machinery, which interferes with host cell signaling. Thereby, effector molecules are introduced or released by the pathogens during the invasion of the host [55, 58, 59]. Autophagy is a evolutionarily refined and sophisticated process wherein molecular cell signaling and cellular metabolism regulation converge to regulate the intermediary metabolism (Figure 3) including the lipid metabolism through a process called lipophagy, which modulate the degradation of lipid droplets in triglycerides and free fatty acid that can be used as a fuel to elevate the rate of β-oxidation and consequently of energy production [60-62], which recently has been demonstrated that some pathogenic agents can subvert this cell process to ensure their own survival.

Recent investigations using genetic, cell biology, and biochemical approaches have led to a better understanding of mechanistic interaction between pathogens and hosts. Based on this, a resource that permit integrate terms of ViralZone, UniProtKB, and GO, has been created, which provide a global view of viral biology and their complex host interactions, based in evasive adaptations and inactivation of antiviral effectors [63]. Advancements in research are

http://dx.doi.org/10.5772/61305

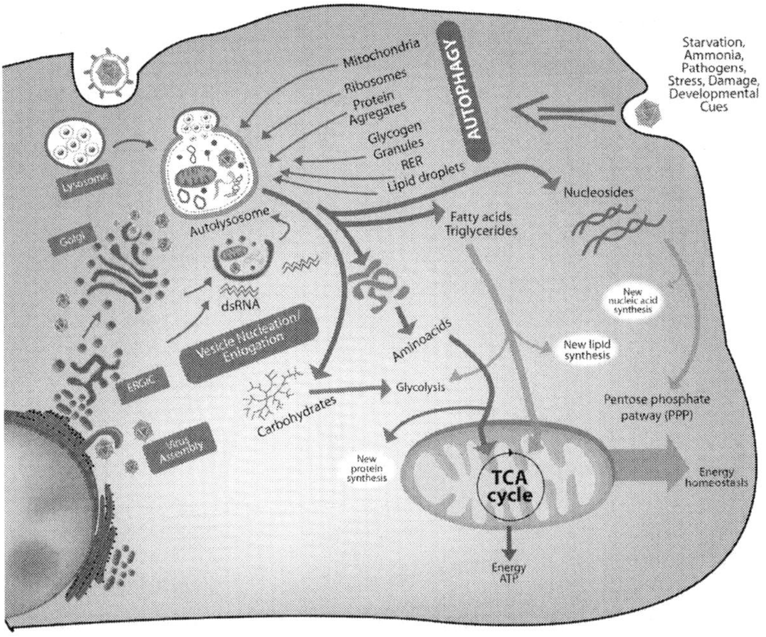

Figure 3. Autophagy and intermediary metabolism relationship

now fueled by increasing interests aimed at the discovery of novel therapeutic interventions against major infectious diseases [64]. The cell biology of microbial pathogenesis has opened many doors for future research into the role of lipids in host-pathogen interactions because lipids of both host and pathogen play critical roles in the pathogen stability to replicate and persist in host cells, and these interactions are very complex and dynamic [58].

The metabolic host cost and contribution of lipids (biosynthesis, catabolism, and trafficking) to the formation of replication factories is in the early stages of investigation [65-69], and yet is need to know pathogen and host lipid profiles as a starting point for tests of functional relevance and comparative profiling in several physiological conditions (status before/after infection) to better understand the details of the metabolic role in the different conditions of the disease and dissect the complicated signaling during host-pathogen interactions for developing drugs and disease biomarkers pathways identification [64].

4. Role of autophagy in infectious disease

The intracellular invaders, after million years of evolution, have developed several sophisticated strategies for evading the host defenses like the immune system. In this regard, autophagy is a complex cellular process that can have a dual role in viral infections depending on the pathogenic agent and host [70, 71]. Although it has been extensively cited and reviewed, the role of autophagy in maintaining the cellular homeostasis [12, 13] still remains to be elucidated in terms of what is their precise role in viral infection.

Considering several infectious agents, there are a number of important findings. For example, macrophages can eliminate *Legionella pneumophila* infection through cholesterol or lipid-raft-rich induction of autophagy [72]. Mycobacteria usurp the host lipid stores for energy production via β- oxidation of fatty acyls, using the glyoxylate cycle enzymes isocitrate lyases for survival and persistence in its human host [73, 74]. *Helicobacter pylori* have been related with elevated cholesterol levels and metabolic syndrome alterations. However, it remains controversial [75, 76]. In HSV-1 (Herpes Simplex Virus), the virulence factor ICP34.5 inhibits autophagy via inhibition of Beclin 1 and PKR [77], and Us3 acts as a viral Akt surrogate to activate mTORC1 inhibiting host autophagy [78]. Curiously, additional members of the herpes virus family employ similar strategies to inhibit autophagy. Gamma herpes virus 68 (gHV68) encodes a virulence factor vbcl2 (M11), which inhibits host autophagy via interaction with Beclin 1 [79]. Kaposi's Sarcoma Herpes Viruses (KSHV) interact with ATG3 and inhibit autophagy [80]. Human Cytomegalovirus (HCMV) inhibits autophagy via upstream activation of mTOR signaling [81]. Autophagy functions as an antiviral host defense of central nervous system against Sindbis Virus (SIN) infection [40, 41].

Hepatitis C virus (HCV) infection has a controversial role in lipid metabolism and autophagy. It has shown that this infection is associated with enhanced lipogenesis, reduced β-oxidation, decreased lipoprotein secretion, and increased autophagy counteracting the alterations in lipid metabolism induced by HCV. In this way, a disruption of autophagic process might contribute to develop steatosis (occurs in about 50% or more of patients) in patients with HCV [82, 83].

It has also been described that the infection of human cells with Poliovirus and Rhinovirus induces autophagosome formation, which are used as sites of viral RNA replication [84]. Autophagosome is required for the formation of Coronavirus replication complexes with the formation of the double membrane vesicles significantly enhancing viral replication efficiency [85]. The use of small interfering RNAs against LC3 or Atg12 has shown to reduce both the intracellular and extracellular yields of poliovirus (+ss) [84]. Reduction in the intracellular concentration of Atg7 reduces the amount of viral capsid protein synthesized in Coxsackievirus B3 [86]. Hepatitis C Virus (HCV) infection was found to activate autophagy, and it extends cell survival for the establishment of a successful viral infection [87].

It should be noted that not only RNA virus (poliovirus, etc.) but also DNA virus (Epstein-Barr virus) infection can induce autophagic machinery, and whether the activation of autophagic machinery can enhance viral replication (poliovirus and mouse hepatitis virus) or not (Vaccinia virus and Herpes Simplex Virus type 1 etc.) depends on the type of viruses [84, 85, 88-90] and on cell type infected [91]. Thus, for some areas for research, the development of the specific inducers of autophagy will offer a promise as a novel class of antiviral therapeutics [16], while for others, the design of specific inhibitors of autophagy could provide new therapeutic strategies [92]. Either will serve as a powerful tool to dissect the autophagic process.

http://dx.doi.org/10.5772/61305

In summary, many different viruses and other pathogens can induce the cellular process, such as apoptosis and autophagy, and on the other hand, host cells can also activate the same pathways when they participate in clearance of infectious agent (Figure 4). Thus, although some viruses may encode one or more inhibitors of both these processes, others have been shown to induce autophagosome-like structures and to benefit from their formation, which may be critical for the viral spread within the infected tissues [35]. Although numerous studies support the beneficial role of autophagy in +ssRNA virus replication [23], the induction of this process is not always favorable for them. And drawing the path and explaining this behavior have shown interesting findings for some researchers.

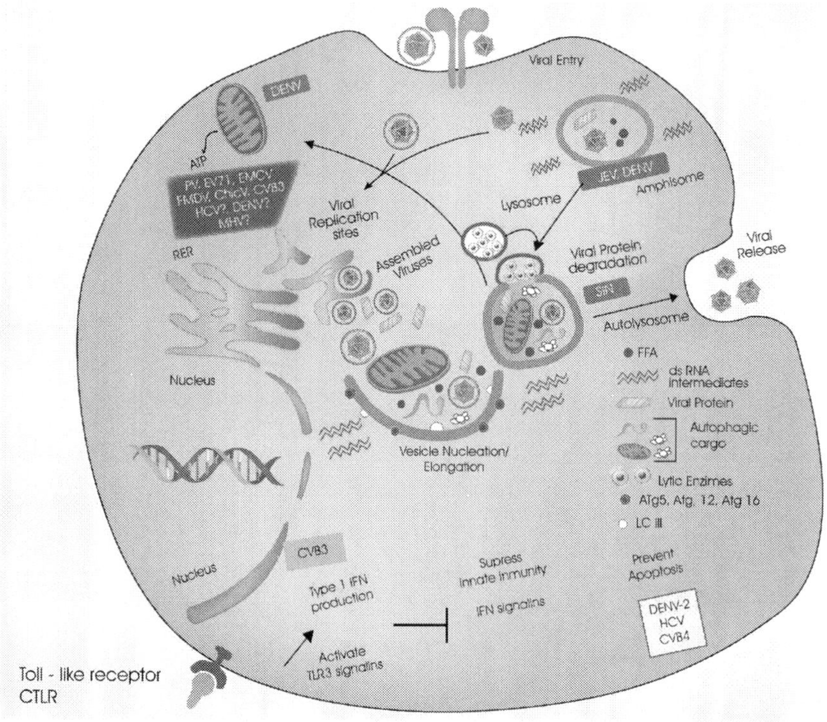

Figure 4. Viral modulation of autophagy. Several viruses have been shown to block or activate various stages of autophagy process.

5. DENV infection and autophagy: Molecular and metabolic convergences

DENV is a major but neglected global public health problem, and despite many efforts, they are made to understand the mechanisms by which it usurps the host cells and this research field has grown dramatically during the last years with multiple studies in molecular and evolutionary biology [93-96], genome sequencing [97-99], construction of infectious clones [100], and use of these to attempt to dissect the specific role of each viral protein [101, 102], and immunological approaches [103, 104]. All of these have failed to produce results that allow the design of vaccines or drugs effective to cure this disease [105-107], and the secrets of DENV and its pathogenesis remain unclear. However, a recent viewpoint of the disease highlights the relevance of the relationship between both hosts and vector systems with the viruses and assign a key role to energy metabolism alterations during the infection, indicating that the virus reprogram the central carbon metabolism (lipid, glucose, TCA cycle and others) [108-112] in order to facilitate their own replication.

Understanding how DENV can differently infect mammals and insect cells is a very interesting issue. In the last years, there has been a notable increase in the research for mosquito DENV infection, revealing the importance of identifying this dual behavior between the host and vector in order to know the cell biology of viral infection. In the enveloped positive-sense RNA viruses such as DENV, a cytoplasmic replication of its are associated with a dramatic rearrangement of host cellular membranes, the merging of viral and target cell membranes, and endosomal trafficking routes are essential to carry out a successful replication cycle [113], and these virus-induced changes the result in induction of vesicular structures that envelope the virus replication complex [114].

A few years ago, it was postulated that autophagosomes might play a structural role in the replication complex formation, and numerous investigations about the role of autophagy in DENV infection were conduced. In 2008, a researcher group from the National Cheng University in Taiwan was the first to demonstrate that DENV can activate autophagic machinery and induce autophagosome formation to promote viral replication, and ATG5 is directly implicated in this activation process [6]. Also, it has been demonstrated that DENV2 induce autophagy and prevent premature cell death, thus, an inhibition of autophagy abolishes its protective role against cell death providing an unfavorable environment for the viral propagation leading to a reduced viral replication [115]. There were experiments to compare single-cycle infections of murine embryonic fibroblasts derived from autophagy-proficient and autophagy-deficient mice showing clear reductions in the yield of extracellular virus in the absence of a functional autophagy pathway [6]. But in 2009, it was demonstrated that DENV replicates on endoplasmic reticulum (ER) cisternae invaginations and not on classical autophagosomes [116]. From this discovery, scientists kept researching the role of autophagy induction in DENV infection.

In the same year, it was shown that the DENV Capsid structural protein contained determinants for lipid droplets targeting. This association was a determinant for reach DENV yield [117], and this discovery was associated with previous findings that reported liver vacuolization and steatosis in DENV infected mice and fatal human cases of DHF [118-120], suggesting a possible role for lipid metabolism in DENV pathogenesis. That was when researchers reported that autophagy process induced by DENV infection plays an indirect role in DENV replication by the modulation of cellular lipid metabolism. Furthermore, it stimulated a cellular triglycerides depletion that are stored in the lipid droplets, leading to release free fatty acids, increased β-oxidation, and energy production to raise the virus yield [121-125] (Figure 5). Subsequently, it was shown that autophagy is mediated in a cell type specific manner, given that autophagy does not have a significant role in DENV replication in monocytic cells [91]. More recent studies in suckling mice demonstrated that DENV infection induce autophagy mechanism *in vivo,* and it played an important role in viral replication, clinical symptoms development, and survival rate [126]. Although it has been widely supported that the autophagy role in DENV infection is more related to a metabolic requirement, it also has been shown that the autophagy pathway plays a determinant role in viral maturation [124], which conduce to think that autophagy does not have a unique function in the viral replication cycle.

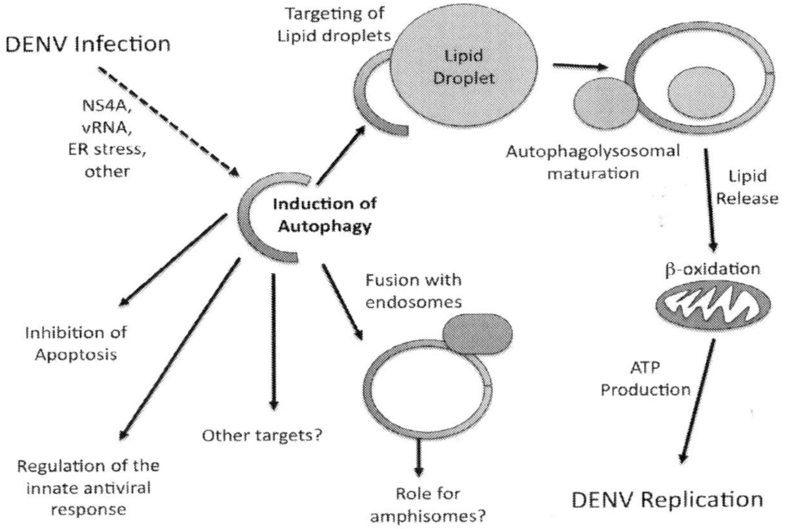

Figure 5. Roles for autophagy during DENV infection From [122].

The discovery of vector factors altered during DENV infection of mosquito may help to identify conserved protein families and pathways that represent both anti-viral mechanisms and requirements for viral life cycle in the vector, and understanding these effects in mosquito vector and correlating it with conserved mammal pathways could help to comprehend the host interactions and development of methods to treat and prevent viral infection and spread. In this way, the mosquito vector, as well as the cell lines, derived from it, was transformed in novel and interesting study models. It has been described in mosquito MAL04 and C6/36 cells that DENV ensures its fusion in late endosomes exploiting cell-controlled differences between lipid compositions of different organelles through interactions between virus and endosomes rich in anionic lipids, protecting against premature release, viral inactivation, or endosome fusion pore opening [127].

The *fat body* plays a major role in intermediary metabolism, and it is the central storage depot of nutrients and energy reserves essential for the holometabolous insects' life, which must accumulate at least a minimal amount of nutrients in larval stages to survive during starvation and metamorphosis. Lipids, mainly triglycerides, represent the major component of *fat body* and are the main source of metabolic fuel, it are stored in the core of lipid droplets, which are mobilized for several purposes as energy provision to flight muscles, ovaries lipids provision and overall maintenance of metabolic activity [128]. The lipolytic machinery identified in insects includes two lipases: TGL and Brummer lipase, and two evolutionarily conserved lipid droplet proteins, Lsd1 and Lsd2 [129]. Current information indicates that insects share with mammals and other organisms, several aspects of the mechanisms of deposition and mobilization of triglycerides. This information validates the use of insect models to investigate basic questions related to the processes of lipid storage and mobilization [130].

DENV drastically alters the lipid profile of mosquito-infected cells, increasing the expression of lipids that have the capacity to change the physical properties of the bilayer such as: bilayer curvature, permeability, and recruitment/assembly of protein complexes in the membrane. Several of the identified molecules also function as bioactive messengers that control signaling and membrane trafficking pathways in the cells [131]. These observations shed light on the emerging role of lipids in shaping the membrane and protein environments during viral infections and suggest membrane organizing principles that may influence virus-induced intracellular membrane architecture [131]. Later, a transcriptome study in *Aedes aegypti* infected with several flaviviruses (WNV, DENV and YFV) was described and an expression profile was observed with 20 significantly upregulated genes and 15 downregulated genes quite similar among them. Something of these genes were related with the regulation of genic expression (juvenile hormone-inducible protein, core histone H3), genes related with antiviral response were downregulated (Jak-STAT pathway downregulated, Toll pathway) and other genes related to ion binding, ion transport, several *metabolic processes* and peptidase activity [132].

In [133], a mosquito protein interaction network based on large-scale protein interaction datasets was developed, and 714 putative dengue-associated mosquito proteins (physical interaction assays, RNAi and microarray) were identified and predicted. Subsequent analysis of these proteins highlighted a sub-network, four regions of highly interconnected proteins with closely related functions (replication/transcription/translation (RTT), immunity, transport, and metabolism). 15 out of 23 proteins (65.2%) were highly interconnected in metabolism region. Consequently, the host infected by the virus can experiment dramatic metabolic alterations. These results support the presence of some common host requirement of DENV in humans and mosquitoes.

6. Concluding remarks and perspectives

After reviewing the historical issues about biochemistry with emphases on metabolism, together with the remarkable findings in cell molecular biology of autophagy pathways, it is clear that right now, we have a great open field for research. Curiously, the animal viruses, during several decades, had been studied, but under the viewpoint of virus-host cell interactions wherein the cell and the viruses have been considered isolated entities. Only in the last decade where the Cell Biology of Virus Infection emerged [134] was considered the cell as the structural and functional unit of infection. Therefore, now, the animal viruses play a role in the physiology of the cell, mimicry and using the metabolic and autophagic cell pathways for the completion of their viral cycles.

An overview was shown here for understanding the viral disease and other human pathologies from an integrative perspective including the theoretical framework and methodology of biochemistry fused to the molecular biology in a cellular compartment (autophagosome), which is triggered for several injuries and/or diseases.

It is important to establish the differences in mechanisms of infection. Therefore, in the basic requirements for this process in both vector (mosquito) and host (mammal), it is important to determine whether it alters in a similar way the metabolism in both models, although the molecular signaling through which these metabolic changes are induced to be different for everyone. It is very interesting that all of these recent researches in mosquitoes suggest alterations in JACK / STAT signaling, toll-like receptors, and metabolism (especially lipid). But knowing that autophagy is conserved from yeast to mammals, the role of autophagy has not been reported in DENV infection in mosquitoes. Moreover, there is a recent research which supports that autophagy is not decisive in the infection in monocytes. It appears that the autophagy is dependent on the cell type.

Together the ideas exposed here with the remarkable findings of several researchers give us a whole landscape where it is possible to find some cellular processes or events, which can be modulated by drugs trying to discover new therapeutical tools.

Acknowledgements

This research was supported by COLCIENCIAS grant 111554531592 from the Colombian government. JCGG was the recipient of a Full-Time Professor Program (Exclusive Dedication) for the Medicine Faculty at the University of Antioquia for 2014–2015.

Author details

Elizabeth Orozco-García and Juan Carlos Gallego-Gómez*

*Address all correspondence to: juanc.gallegomez@gmail.com

Translational and Molecular Medicine Group, Medellín Medical Research Institute, Faculty of Medicine, Universidad de Antioquia, Medellín, Colombia

References

[1] Solomon T, Mallewa M. 2001. Dengue and other emerging flaviviruses. J Infect 42:104-115.

[2] Guzman MG, Halstead SB, Artsob H, Buchy P, Farrar J, Gubler DJ, Hunsperger E, Kroeger A, Margolis HS, Martinez E, Nathan MB, Pelegrino JL, Simmons C, Yoksan S, Peeling RW. 2010. Dengue: a continuing global threat. Nat Rev Microbiol 8:S7-16.

[3] Bhatt S, Gething PW, Brady OJ, Messina JP, Farlow AW, Moyes CL, Drake JM, Brownstein JS, Hoen AG, Sankoh O, Myers MF, George DB, Jaenisch T, Wint GR, Simmons CP, Scott TW, Farrar JJ, Hay SI. 2013. The global distribution and burden of dengue. Nature 496:504-507.

[4] Gubler DJ. 2012. The Economic Burden of Dengue. Am J Trop Med Hyg 86:743-744.

[5] Low JGH, Ong A, Tan LK, Chaterji S, Chow A, Lim WY, Lee KW, Chua R, Chua CR, Tan SWS, Cheung YB, Hibberd ML, Vasudevan SG, Ng LC, Leo YS, Ooi EE. 2011. The Early Clinical Features of Dengue in Adults: Challenges for Early Clinical Diagnosis, PLoS Negl Trop Dis, vol 5.

[6] Lee YR, Lei HY, Liu MT, Wang JR, Chen SH, Jiang-Shieh YF, Lin YS, Yeh TM, Liu CC, Liu HS. 2008. Autophagic machinery activated by dengue virus enhances virus replication. Virology 374:240-248.

[7] Gubler DJ. 1998. Dengue and dengue hemorrhagic fever. Clin Microbiol Rev 11:480-496.

[8] Kurane I. 2007. Dengue hemorrhagic fever with special emphasis on immunopathogenesis. Comp Immunol Microbiol Infect Dis 30:329-340.

[9] Lindenbach BD, H.-J. Thiel, C.M. Rice. 2007. "Flaviviridae: The Viruses and Their Replication." *In* Howley DMKaPM (ed), Fields Virology, 5th Edition ed. Lippincott-Raven Publishers.

[10] Black WCt, Bennett KE, Gorrochotegui-Escalante N, Barillas-Mury CV, Fernandez-Salas I, de Lourdes Munoz M, Farfan-Ale JA, Olson KE, Beaty BJ. 2002. Flavivirus susceptibility in Aedes aegypti. Arch Med Res 33:379-388.

[11] Bäck AT, Lundkvist Å. 2013. Dengue viruses – an overview. Infect Ecol Epidemiol 3:10.3402/iee.v3403i3400.19839.

[12] Ravikumar B, Sarkar S, Davies JE, Futter M, Garcia-Arencibia M, Green-Thompson ZW, Jimenez-Sanchez M, Korolchuk VI, Lichtenberg M, Luo S, Massey DC, Menzies FM, Moreau K, Narayanan U, Renna M, Siddiqi FH, Underwood BR, Winslow AR, Rubinsztein DC. 2010. Regulation of mammalian autophagy in physiology and pathophysiology. Physiol Rev 90:1383-1435.

[13] Levine B, Kroemer G. 2008. Autophagy in the pathogenesis of disease. Cell 132:27-42.

[14] Mizushima N, Levine B, Cuervo AM, Klionsky DJ. 2008. Autophagy fights disease through cellular self-digestion. Nature 451:1069-1075.

[15] Klionsky DJ. 2007. Autophagy: from phenomenology to molecular understanding in less than a decade. Nat Rev Mol Cell Biol 8:931-937.

[16] Yordy B, Iwasaki A. 2011. Autophagy in the control and pathogenesis of viral infection. Curr Opin Virol 1:196-203.

[17] Kelekar A. 2006. Autophagy. Annals of the New York Academy of Sciences 1066:259-271.

[18] Levine B, Mizushima N, Virgin HW. 2011. Autophagy in immunity and inflammation. Nature 469:323-335.

[19] Djavaheri-Mergny M, Maiuri MC, Kroemer G. 2010. Cross talk between apoptosis and autophagy by caspase-mediated cleavage of Beclin 1. Oncogene 29:1717-1719.

[20] Levine B, Sinha S, Kroemer G. 2008. Bcl-2 family members: dual regulators of apoptosis and autophagy. Autophagy 4:600-606.

[21] Maiuri MC, Zalckvar E, Kimchi A, Kroemer G. 2007. Self-eating and self-killing: crosstalk between autophagy and apoptosis. Nat Rev Mol Cell Biol 8:741-752.

[22] Mariño G, Niso-Santano M, Baehrecke EH, Kroemer G. 2014. Self-consumption: the interplay of autophagy and apoptosis. Nat Rev Mol Cell Biol 15:81-94.

[23] Shi J, Luo H. 2012. Interplay between the cellular autophagy machinery and positive-stranded RNA viruses. Acta Biochim Biophys Sin (Shanghai) 44:375-384.

[24] Kamada Y, Funakoshi T, Shintani T, Nagano K, Ohsumi M, Ohsumi Y. 2000. Tor-mediated induction of autophagy via an Apg1 protein kinase complex. J Cell Biol 150:1507-1513.

[25] Suzuki K, Kubota Y, Sekito T, Ohsumi Y. 2007. Hierarchy of Atg proteins in pre-autophagosomal structure organization. Genes Cells 12:209-218.

[26] Epple UD, Suriapranata I, Eskelinen EL, Thumm M. 2001. Aut5/Cvt17p, a putative lipase essential for disintegration of autophagic bodies inside the vacuole. J Bacteriol 183:5942-5955.

[27] Teter SA, Eggerton KP, Scott SV, Kim J, Fischer AM, Klionsky DJ. 2001. Degradation of lipid vesicles in the yeast vacuole requires function of Cvt17, a putative lipase. J Biol Chem 276:2083-2087.

[28] Tanida I. 2011. Autophagy basics. Microbiol Immunol 55:1-11.

[29] Stroupe C. 2011. Autophagy: cells SNARE selves. Curr Biol 21:R697-699.

[30] Chua CE, Gan BQ, Tang BL. 2011. Involvement of members of the Rab family and related small GTPases in autophagosome formation and maturation. Cell Mol Life Sci 68:3349-3358.

[31] He C, Klionsky DJ. 2009. Regulation mechanisms and signaling pathways of autophagy. Annu Rev Genet 43:67-93.

[32] Jager S, Bucci C, Tanida I, Ueno T, Kominami E, Saftig P, Eskelinen EL. 2004. Role for Rab7 in maturation of late autophagic vacuoles. J Cell Sci 117:4837-4848.

[33] Sahu R, Kaushik S, Clement CC, Cannizzo ES, Scharf B, Follenzi A, Potolicchio I, Nieves E, Cuervo AM, Santambrogio L. 2011. Microautophagy of cytosolic proteins by late endosomes. Dev Cell 20:131-139.

[34] Cuervo AM. 2010. Chaperone-mediated autophagy: selectivity pays off. Trends Endocrinol Metab 21:142-150.

[35] Kirkegaard K. 2009. Subversion of the cellular autophagy pathway by viruses. Curr Top Microbiol Immunol 335:323-333.

[36] Kudchodkar SB, Levine B. 2009. Viruses and autophagy. Rev Med Virol 19:359-378.

[37] Lee HK, Iwasaki A. 2008. Autophagy and antiviral immunity. Curr Opin Immunol 20:23-29.

[38] Orvedahl A, Levine B. 2009. Autophagy in Mammalian antiviral immunity. Curr Top Microbiol Immunol 335:267-285.

[39] Taylor MP, Jackson WT. 2009. Viruses and arrested autophagosome development. Autophagy 5:870-871.

[40] Liang XH, Kleeman LK, Jiang HH, Gordon G, Goldman JE, Berry G, Herman B, Levine B. 1998. Protection against fatal Sindbis virus encephalitis by beclin, a novel Bcl-2-interacting protein. J Virol 72:8586-8596.

[41] Orvedahl A, MacPherson S, Sumpter R, Jr., Talloczy Z, Zou Z, Levine B. 2010. Autophagy protects against Sindbis virus infection of the central nervous system. Cell Host Microbe 7:115-127.

[42] Tallóczy Z, Jiang W, Virgin HW, Leib DA, Scheuner D, Kaufman RJ, Eskelinen EL, Levine B. 2002. Regulation of starvation- and virus-induced autophagy by the eIF2α kinase signaling pathway. Proc Natl Acad Sci U S A 99:190-195.

[43] DeBerardinis RJ, Thompson CB. 2012. Cellular metabolism and disease: what do metabolic outliers teach us? Cell 148:1132-1144.

[44] Hanson RW. 2005. Metabolism in the era of molecular biology. J Biol Chem 280:1705-1715.

[45] McKnight SL. 2010. On getting there from here. Science 330:1338-1339.

[46] Takahashi H, McCaffery JM, Irizarry RA, Boeke JD. 2006. Nucleocytosolic acetyl-coenzyme a synthetase is required for histone acetylation and global transcription. Mol Cell 23:207-217.

[47] Wellen KE, Hatzivassiliou G, Sachdeva UM, Bui TV, Cross JR, Thompson CB. 2009. ATP-citrate lyase links cellular metabolism to histone acetylation. Science 324:1076-1080.

[48] Zhao S, Xu W, Jiang W, Yu W, Lin Y, Zhang T, Yao J, Zhou L, Zeng Y, Li H, Li Y, Shi J, An W, Hancock SM, He F, Qin L, Chin J, Yang P, Chen X, Lei Q, Xiong Y, Guan KL. 2010. Regulation of cellular metabolism by protein lysine acetylation. Science 327:1000-1004.

[49] Sommer P, Sweeney G. 2010. Functional and Mechanistic Integration of Infection and the Metabolic Syndrome. Korean Diabetes J 34:71-76.

[50] DeBerardinis RJ, Lum JJ, Hatzivassiliou G, Thompson CB. 2008. The biology of cancer: metabolic reprogramming fuels cell growth and proliferation. Cell Metab 7:11-20.

[51] Deberardinis RJ, Sayed N, Ditsworth D, Thompson CB. 2008. Brick by brick: metabolism and tumor cell growth. Curr Opin Genet Dev 18:54-61.

[52] DeBerardinis RJ. 2008. Is cancer a disease of abnormal cellular metabolism? New angles on an old idea. Genet Med 10:767-777.

[53] DeBerardinis RJ. 2010. 2010 Keystone Symposium: Metabolism and Cancer Progression. Future Oncol 6:893-895.

[54] Butel JS. 2000. Viral carcinogenesis: revelation of molecular mechanisms and etiology of human disease. Carcinogenesis 21:405-426.

[55] Alto NM, Orth K. 2012. Subversion of cell signaling by pathogens. Cold Spring Harb Perspect Biol 4:a006114.

[56] Meckes DG, Jr., Raab-Traub N. 2011. Microvesicles and viral infection. J Virol 85:12844-12854.

[57] Ali N, Allam H, May R, Sureban SM, Bronze MS, Bader T, Umar S, Anant S, Houchen CW. 2011. Hepatitis C virus-induced cancer stem cell-like signatures in cell culture and murine tumor xenografts. J Virol 85:12292-12303.

[58] Wenk MR. 2006. Lipidomics of host-pathogen interactions. FEBS Lett 580:5541-5551.

[59] Alomairi J, Cellular Stress CdReCdM, INSERM UMR 1068, CNRS UMR 7258, Aix-Marseille University and Institut Paoli-Calmettes, Marseille, France, Infections GaPL, URMITE-IRD198, INSERM U1095, CNRS UMR7278, Aix-Marseille University, Marseille, France, Bonacci T, Cellular Stress CdReCdM, INSERM UMR 1068, CNRS UMR 7258, Aix-Marseille University and Institut Paoli-Calmettes, Marseille, France, Ghigo E, Infections GaPL, URMITE-IRD198, INSERM U1095, CNRS UMR7278, Aix-Marseille University, Marseille, France, Soubeyran P, Cellular Stress CdReCdM, INSERM

UMR 1068, CNRS UMR 7258, Aix-Marseille University and Institut Paoli-Calmettes, Marseille, France, philippe.soubeyran@inserm.fr. 2015. Alterations of host cell ubiquitination machinery by pathogenic bacteria. Frontiers in Cellular and Infection Microbiology 5:7.

[60] Dong H, Czaja MJ. 2011. Regulation of lipid droplets by autophagy. Trends Endocrinol Metab 22:234-240.

[61] Singh R, Kaushik S, Wang Y, Xiang Y, Novak I, Komatsu M, Tanaka K, Cuervo AM, Czaja MJ. 2009. Autophagy regulates lipid metabolism. Nature 458:1131-1135.

[62] Liu K, Czaja MJ. 2013. Regulation of lipid stores and metabolism by lipophagy. Cell Death Differ 20:3-11.

[63] Masson P, Hulo C, de Castro E, Foulger R, Poux S, Bridge A, Lomax J, Bougueleret L, Xenarios I, Le Mercier P. 2014. An integrated ontology resource to explore and study host-virus relationships. PLoS One 9:e108075.

[64] Wenk MR. 2005. The emerging field of lipidomics. Nat Rev Drug Discov 4:594-610.

[65] Alvisi G, Madan V, Bartenschlager R. 2011. Hepatitis C virus and host cell lipids: an intimate connection. RNA Biol 8:258-269.

[66] Diamond DL, Syder AJ, Jacobs JM, Sorensen CM, Walters KA, Proll SC, McDermott JE, Gritsenko MA, Zhang Q, Zhao R, Metz TO, Camp DG, 2nd, Waters KM, Smith RD, Rice CM, Katze MG. 2010. Temporal proteome and lipidome profiles reveal hepatitis C virus-associated reprogramming of hepatocellular metabolism and bioenergetics. PLoS Pathog 6:e1000719.

[67] Munger J, Bennett BD, Parikh A, Feng XJ, McArdle J, Rabitz HA, Shenk T, Rabinowitz JD. 2008. Systems-level metabolic flux profiling identifies fatty acid synthesis as a target for antiviral therapy. Nat Biotechnol 26:1179-1186.

[68] Spencer CM, Schafer XL, Moorman NJ, Munger J. 2011. Human cytomegalovirus induces the activity and expression of acetyl-coenzyme A carboxylase, a fatty acid biosynthetic enzyme whose inhibition attenuates viral replication. J Virol 85:5814-5824.

[69] Wang X, Diaz A, Hao L, Gancarz B, den Boon JA, Ahlquist P. 2011. Intersection of the multivesicular body pathway and lipid homeostasis in RNA replication by a positive-strand RNA virus. J Virol 85:5494-5503.

[70] Chiramel AI, Brady NR, Bartenschlager R. 2013. Divergent Roles of Autophagy in Virus Infection. Cells 2:83-104.

[71] Dong X, Levine B. 2013. Autophagy and viruses: adversaries or allies? J Innate Immun 5:480-493.

[72] Amer AO, Swanson MS. 2005. Autophagy is an immediate macrophage response to Legionella pneumophila. Cell Microbiol 7:765-778.

[73] Munoz-Elias EJ, McKinney JD. 2005. Mycobacterium tuberculosis isocitrate lyases 1 and 2 are jointly required for in vivo growth and virulence. Nat Med 11:638-644.

[74] Lee W, VanderVen BC, Fahey RJ, Russell DG. 2013. Intracellular Mycobacterium tuberculosis exploits host-derived fatty acids to limit metabolic stress. J Biol Chem 288:6788-6800.

[75] Albaker WI. 2011. Helicobacter pylori Infection and its Relationship to Metabolic Syndrome: Is it a Myth or Fact? Saudi J Gastroenterol 17:165-169.

[76] Feingold KR, Grunfeld C. 2012. Lipids: a key player in the battle between the host and microorganisms. J Lipid Res 53:2487-2489.

[77] Orvedahl A, Alexander D, Talloczy Z, Sun Q, Wei Y, Zhang W, Burns D, Leib DA, Levine B. 2007. HSV-1 ICP34.5 confers neurovirulence by targeting the Beclin 1 autophagy protein. Cell Host Microbe 1:23-35.

[78] Chuluunbaatar U, Roller R, Feldman ME, Brown S, Shokat KM, Mohr I. 2010. Constitutive mTORC1 activation by a herpesvirus Akt surrogate stimulates mRNA translation and viral replication. Genes Dev 24:2627-2639.

[79] Ku B, Woo JS, Liang C, Lee KH, Hong HS, E X, Kim KS, Jung JU, Oh BH. 2008. Structural and biochemical bases for the inhibition of autophagy and apoptosis by viral BCL-2 of murine gamma-herpesvirus 68. PLoS Pathog 4:e25.

[80] Lee JS, Li Q, Lee JY, Lee SH, Jeong JH, Lee HR, Chang H, Zhou FC, Gao SJ, Liang C, Jung JU. 2009. FLIP-mediated autophagy regulation in cell death control. Nat Cell Biol 11:1355-1362.

[81] Chaumorcel M, Souquere S, Pierron G, Codogno P, Esclatine A. 2008. Human cytomegalovirus controls a new autophagy-dependent cellular antiviral defense mechanism. Autophagy 4:46-53.

[82] Syed GH, Amako Y, Siddiqui A. 2010. Hepatitis C virus hijacks host lipid metabolism. Trends Endocrinol Metab 21:33-40.

[83] Vescovo T, Romagnoli A, Perdomo AB, Corazzari M, Ciccosanti F, Alonzi T, Nardacci R, Ippolito G, Tripodi M, Garcia-Monzon C, Lo Iacono O, Piacentini M, Fimia GM. 2012. Autophagy protects cells from HCV-induced defects in lipid metabolism. Gastroenterology 142:644-653.e643.

[84] Jackson WT, Giddings TH, Jr., Taylor MP, Mulinyawe S, Rabinovitch M, Kopito RR, Kirkegaard K. 2005. Subversion of cellular autophagosomal machinery by RNA viruses. PLoS Biol 3:e156.

[85] Prentice E, Jerome WG, Yoshimori T, Mizushima N, Denison MR. 2004. Coronavirus replication complex formation utilizes components of cellular autophagy. J Biol Chem 279:10136-10141.

[86] Wong J, Zhang J, Si X, Gao G, Mao I, McManus BM, Luo H. 2008. Autophagosome supports coxsackievirus B3 replication in host cells. J Virol 82:9143-9153.

[87] Shrivastava S, Raychoudhuri A, Steele R, Ray R, Ray RB. 2011. Knockdown of autophagy enhances the innate immune response in hepatitis C virus-infected hepatocytes. Hepatology 53:406-414.

[88] Alexander DE, Ward SL, Mizushima N, Levine B, Leib DA. 2007. Analysis of the role of autophagy in replication of herpes simplex virus in cell culture. J Virol 81:12128-12134.

[89] Wileman T. 2006. Aggresomes and autophagy generate sites for virus replication. Science 312:875-878.

[90] Zhang H, Monken CE, Zhang Y, Lenard J, Mizushima N, Lattime EC, Jin S. 2006. Cellular autophagy machinery is not required for vaccinia virus replication and maturation. Autophagy 2:91-95.

[91] Panyasrivanit M, Greenwood MP, Murphy D, Isidoro C, Auewarakul P, Smith DR. 2011. Induced autophagy reduces virus output in dengue infected monocytic cells. Virology 418:74-84.

[92] Rubinsztein DC, Codogno P, Levine B. 2012. Autophagy modulation as a potential therapeutic target for diverse diseases. Nat Rev Drug Discov 11:709-730.

[93] Usme-Ciro JA, Mendez JA, Tenorio A, Rey GJ, Domingo C, Gallego-Gomez JC. 2008. Simultaneous circulation of genotypes I and III of dengue virus 3 in Colombia. Virol J 5:101.

[94] Mendez JA, Usme-Ciro JA, Domingo C, Rey GJ, Sanchez JA, Tenorio A, Gallego-Gomez JC. 2012. Phylogenetic reconstruction of dengue virus type 2 in Colombia. Virol J 9:64.

[95] Mendez JA, Usme-Ciro JA, Domingo C, Rey GJ, Sanchez JA, Tenorio A, Gallego-Gomez JC. 2010. Phylogenetic history demonstrates two different lineages of dengue type 1 virus in Colombia. Virol J 7:226.

[96] Bartenschlager R, Miller S. 2008. Molecular aspects of Dengue virus replication. Future Microbiol 3:155-165.

[97] Carter JR, Keith JH, Barde PV, Fraser TS, Fraser MJ, Jr. 2010. Targeting of highly conserved Dengue virus sequences with anti-Dengue virus trans-splicing group I introns. BMC Mol Biol 11:84.

[98] Zhao H, Deng YQ, Hong WX, Yu XD, Jiang T, Yu M, Hu FY, Zhu SY, Li XF, Song KY, Qin ED, Zhang FC, Qin CF. 2012. Complete genome sequence of dengue virus serotype 2 Cosmopolitan genotype strain in Guangdong, China. J Virol 86:13808-13809.

[99] Barrero PR, Mistchenko AS. 2004. Complete genome sequencing of dengue virus type 1 isolated in Buenos Aires, Argentina. Virus Res 101:135-145.

[100] Usme-Ciro JA, Lopera JA, Enjuanes L, Almazan F, Gallego-Gomez JC. 2014. Development of a novel DNA-launched dengue virus type 2 infectious clone assembled in a bacterial artificial chromosome. Virus Res 180:12-22.

[101] Hannemann H, Sung PY, Chiu HC, Yousuf A, Bird J, Lim SP, Davidson AD. 2013. Serotype-specific differences in dengue virus non-structural protein 5 nuclear localization. J Biol Chem 288:22621-22635.

[102] de Wispelaere M, Yang PL. 2012. Mutagenesis of the DI/DIII Linker in Dengue Virus Envelope Protein Impairs Viral Particle Assembly. J Virol 86:7072-7083.

[103] Lei HY, Yeh TM, Liu HS, Lin YS, Chen SH, Liu CC. 2001. Immunopathogenesis of dengue virus infection. J Biomed Sci 8:377-388.

[104] Martina BE, Koraka P, Osterhaus AD. 2009. Dengue virus pathogenesis: an integrated view. Clin Microbiol Rev 22:564-581.

[105] Usme-Ciro JA, Mendez JA, Laiton KD, Paez A. 2014. The relevance of dengue virus genotypes surveillance at country level before vaccine approval. Hum Vaccin Immunother 10:2674-2678.

[106] Thomas SJ, Endy TP. 2011. Critical issues in dengue vaccine development. Curr Opin Infect Dis 24:442-450.

[107] Slifka MK. 2014. Vaccine-Mediated Immunity Against Dengue and the Potential for Long-Term Protection Against Disease. Front Immunol 5.

[108] Pando-Robles V, Oses-Prieto JA, Rodriguez-Gandarilla M, Meneses-Romero E, Burlingame AL, Batista CV. 2014. Quantitative proteomic analysis of Huh-7 cells infected with Dengue virus by label-free LC-MS. J Proteomics 111:16-29.

[109] Birungi G, Chen SM, Loy BP, Ng ML, Li SF. 2010. Metabolomics approach for investigation of effects of dengue virus infection using the EA.hy926 cell line. J Proteome Res 9:6523-6534.

[110] Fontaine KA, Sanchez EL, Camarda R, Lagunoff M. 2014. Dengue Virus Induces and Requires Glycolysis for Optimal Replication. J Virol.

[111] Arnold PA, Johnson KN, White CR. 2013. Physiological and metabolic consequences of viral infection in Drosophila melanogaster. J Exp Biol 216:3350-3357.

[112] Maynard ND, Gutschow MV, Birch EW, Covert MW. 2010. The Virus as Metabolic Engineer. Biotechnol J 5:686-694.

[113] Joo KI, Tai A, Lee CL, Wong C, Wang P. 2010. Imaging multiple intermediates of single-virus membrane fusion mediated by distinct fusion proteins. Microsc Res Tech 73:886-900.

[114] Gillespie LK, Hoenen A, Morgan G, Mackenzie JM. 2010. The endoplasmic reticulum provides the membrane platform for biogenesis of the flavivirus replication complex. J Virol 84:10438-10447.

[115] McLean JE, Wudzinska A, Datan E, Quaglino D, Zakeri Z. 2011. Flavivirus NS4A-induced autophagy protects cells against death and enhances virus replication. J Biol Chem 286:22147-22159.

[116] Welsch S, Miller S, Romero-Brey I, Merz A, Bleck CK, Walther P, Fuller SD, Antony C, Krijnse-Locker J, Bartenschlager R. 2009. Composition and three-dimensional architecture of the dengue virus replication and assembly sites. Cell Host Microbe 5:365-375.

[117] Samsa MM, Mondotte JA, Iglesias NG, Assuncao-Miranda I, Barbosa-Lima G, Da Poian AT, Bozza PT, Gamarnik AV. 2009. Dengue virus capsid protein usurps lipid droplets for viral particle formation. PLoS Pathog 5:e1000632.

[118] Paes MV, Pinhao AT, Barreto DF, Costa SM, Oliveira MP, Nogueira AC, Takiya CM, Farias-Filho JC, Schatzmayr HG, Alves AM, Barth OM. 2005. Liver injury and viremia in mice infected with dengue-2 virus. Virology 338:236-246.

[119] Huerre MR, Lan NT, Marianneau P, Hue NB, Khun H, Hung NT, Khen NT, Drouet MT, Huong VT, Ha DQ, Buisson Y, Deubel V. 2001. Liver histopathology and biological correlates in five cases of fatal dengue fever in Vietnamese children. Virchows Arch 438:107-115.

[120] Povoa TF, Alves AM, Oliveira CA, Nuovo GJ, Chagas VL, Paes MV. 2014. The pathology of severe dengue in multiple organs of human fatal cases: histopathology, ultrastructure and virus replication. PLoS One 9:e83386.

[121] Heaton NS, Randall G. 2011. Multifaceted roles for lipids in viral infection. Trends Microbiol 19:368-375.

[122] Heaton NS, Randall G. 2011. Dengue Virus and Autophagy. Viruses 3:1332-1341.

[123] Heaton NS, Randall G. 2010. Dengue virus-induced autophagy regulates lipid metabolism. Cell Host Microbe 8:422-432.

[124] Mateo R, Nagamine CM, Spagnolo J, Mendez E, Rahe M, Gale M, Jr., Yuan J, Kirkegaard K. 2013. Inhibition of cellular autophagy deranges dengue virion maturation. J Virol 87:1312-1321.

[125] Fang YT, Wan SW, Lu YT, Yao JH, Lin CF, Hsu LJ, Brown MG, Marshall JS, Anderson R, Lin YS. 2014. Autophagy facilitates antibody-enhanced dengue virus infection in human pre-basophil/mast cells. PLoS One 9:e110655.

[126] Lee YR, Hu HY, Kuo SH, Lei HY, Lin YS, Yeh TM, Liu CC, Liu HS. 2013. Dengue virus infection induces autophagy: an in vivo study. J Biomed Sci 20:65.

[127] Zaitseva E, Yang ST, Melikov K, Pourmal S, Chernomordik LV. 2010. Dengue Virus Ensures Its Fusion in Late Endosomes Using Compartment-Specific Lipids. PLoS Pathog 6.

[128] Arrese EL, Soulages JL. 2010. Insect fat body: energy, metabolism, and regulation. Annu Rev Entomol 55:207-225.

[129] Arrese EL, Patel RT, Soulages JL. 2006. The main triglyceride-lipase from the insect fat body is an active phospholipase A(1): identification and characterization. J Lipid Res 47:2656-2667.

[130] Cheon HM, Shin SW, Bian G, Park JH, Raikhel AS. 2006. Regulation of lipid metabolism genes, lipid carrier protein lipophorin, and its receptor during immune challenge in the mosquito Aedes aegypti. J Biol Chem 281:8426-8435.

[131] Perera R, Riley C, Isaac G, Hopf-Jannasch AS, Moore RJ, Weitz KW, Pasa-Tolic L, Metz TO, Adamec J, Kuhn RJ. 2012. Dengue virus infection perturbs lipid homeostasis in infected mosquito cells. PLoS Pathog 8:e1002584.

[132] Colpitts TM, Cox J, Vanlandingham DL, Feitosa FM, Cheng G, Kurscheid S, Wang P, Krishnan MN, Higgs S, Fikrig E. 2011. Alterations in the Aedes aegypti transcriptome during infection with West Nile, dengue and yellow fever viruses. PLoS Pathog 7:e1002189.

[133] Guo X, Xu Y, Bian G, Pike AD, Xie Y, Xi Z. 2010. Response of the mosquito protein interaction network to dengue infection. BMC Genomics 11:380.

[134] Krijnse-Locker J, Sodeik B, Suomalainen M. 2002. Meeting report from the EMBO workshop "The Cell Biology of Virus Infection", Heidelberg, Germany, 22-26 September 2001. Traffic 3:233-235.

Chapter 7

Macrophages – The Key Actors in Adipose Tissue Remodeling and Dysfunction

Sanja Stojanović and Stevo Najman

Additional information is available at the end of the chapter

http://dx.doi.org/10.5772/62152

Abstract

Adipose tissue (AT) is a very important endocrine and paracrine organ that regulates other tissues and organs. Dysfunction of AT leads to a wide range of disorders like obesity, insulin resistance, diabetes mellitus, cardiac disorders, tumors and others. Adipose tissue macrophages (ATMs) are the key actors in AT remodeling and dysfunction. Their role in AT dysfunction is nowadays increasingly investigated, but still their interplay and molecular mechanisms of actions have not been fully elucidated. In this chapter, we summarized the current knowledge about the role of macrophages in AT remodeling, dysfunction and related disorders and indicate the potential directions for future research.

Keywords: Adipose tissue, macrophages, tissue remodeling, adipose tissue dysfunctions

1. Introduction

Adipose tissue (AT) was previously considered to be only a fat depot. Today, it is well known that AT secretes a large number of proteins collectively termed as adipokines (adiponectin, leptin, resistin and inflammatory cytokines TNF-α, IL6, IL8, IL1, IL10, IL18 and TGF-β) that are responsible for many different processes in the body. Therefore, AT is considered to be a highly active metabolic, endocrine and paracrine organ that regulates other tissues and organs. AT is very heterogeneous and consists of different cell types such as: adipocytes, pre-adipocytes, endothelial cells, fibroblasts, mesenchymal stem cells and immune cells (mast cells, lymphocytes and macrophages). Adipose tissue macrophages (ATMs) are cells that are responsible for AT remodeling. There are two types of ATMs, M1 (classically activated) or inflammatory macrophages and M2 (alternatively activated), anti-inflammatory or reparatory macrophages. The role of ATMs in disorders such as obesity, insulin resistance, diabetes mellitus, cardiac disorders, tumors and others is nowadays increasingly investigated, but still their interplay and molecular mechanisms of actions have not been fully elucidated. This chapter provides an overview of current knowledge about the role of macrophages in AT remodeling, dysfunction and related disorders and indicates the potential directions for future research.

2. Adipose Tissue Macrophages (ATMs)

Although adipocytes play a central role in adipose tissue (AT) remodeling, an increasing attention is directed toward adipose tissue macrophages (ATMs). Since adipose tissue remodeling is nowadays considered as chronic inflammation, ATMs and their interaction with adipocytes are key events that orchestrate the adipose tissue remodeling process.

Resident ATMs are very heterogenic population of cells that is reflected on their function in AT [1, 2]. During AT remodeling, factors that are released from AT induce the recruitment of monocytes into AT. It has been shown that most of the macrophages in AT are derived from bone marrow [3, 4].

There are two types of ATMs: M1 (classically activated) and M2 (alternatively activated) macrophages. They are characterized based on their polarization state, the expression of particular antigens [2, 5, 6] and secretion products. M1 (classically activated) macrophages, also called pro-inflammatory macrophages, are dominant type of macrophages during AT expansion and inflammation. They are characteristic of obese AT. Classically activated macrophages can be induced by LPS and the Th1 cytokine IFN-γ and express high levels of pro-inflammatory mediators including F4/80, CD11c, TNF-α, IL-6, iNOS, CCR2, IL-12 and IL-23 [6–9]. M2 (alternatively activated) macrophages, also called reparatory or anti-inflammatory macrophages, are dominant in lean AT. M2 macrophages are responsible for AT homeostasis, tissue repair and immunosuppression. Exposure of macrophages to the TH2 cytokine IL-4 produces M2 phenotype. They express F4/80, CD301, arginase 1 [6, 7] and CD163 and high levels of scavenger, mannose, and galactose-type receptors. They secrete anti-inflammatory cytokines such as IL-10 and IL-1 receptor antagonist [8, 9] and are shown to inhibit NOS (iNOS) activity. M2 macrophages preserve normal adipocyte function by promoting tissue repair and angiogenesis in an increasing AT mass [2, 10].

3. The role of ATMs in adipose tissue dysfunction and related disorders

The exact role of ATMs in AT dysfunction and related disorders is still not known. In recent years, a lot of research has been done, and it has been shown that the balance between M1 and M2 macrophages is crucial for maintaining normal adipocyte function and AT homeostasis.

Obesity is a very common chronic disease that leads to the development of insulin resistance, diabetes mellitus, cardiac disorders and others [3, 4, 11–15]. Obesity is characterized as a low-grade chronic inflammation with unbalanced production of pro- and anti-inflammatory adipokines that contributes to the development of metabolic syndrome [4, 11–14, 16] and may be involved in a variety of physiologic and pathologic processes [17]. In obesity, the balance between M1 and M2 macrophages is disturbed and moved toward M1 inflammatory macrophages. There are two mechanisms of imbalance occurrence: infiltration of monocytes from circulation under the influence of molecules secreted from growing AT and "phenotypic switching" between M1 and M2 macrophages. During the AT growth, adipocytes secrete products that promote the production of macrophage inflammatory cytokines [18, 19]. These products influence the polarization of resident macrophages. A model of "phenotypic switching" of macrophages has been reported by Lumeng et al. in 2007 [6]. Their model emphasized that obesity is accompanied by a transformation in the polarized states of macrophages, from an "alternatively activated" M2 that primarily accumulates during negative energy balance to a more pro-inflammatory "classically activated" M1 macrophages. This phenotypic change from M2 to M1 polarization in obese adipose tissue leads to adipose tissue inflammation [20–23]. Macrophages that are infiltrated into AT from circulation are an important source of inflammation in obese AT. Chemokines are small pro-inflammatory molecules that promote macrophage mobilization from bone marrow into tissues. Increased expression of chemokines in obese adipose tissue has been implicated in the control of monocyte recruitment to the adipose tissue. During the expansion of AT, secretion of pro-inflammatory cytokines is upregulated and they are released into the circulation. It is shown that MCP-1/CCR2 pathways have pathophysiological role in macrophage infiltration into obese adipose tissue [24, 25]. MCP-1 plays a role in the recruitment of macrophages into obese adipose tissue. Increased levels of MCP-1, CXCL14, MIP-1α, MCP-2, MCP-3 and RANTES can be observed in AT of mice with genetic or DIO [15, 26]. CCR2 expressed in bone marrow cells is involved in macrophage infiltration into obese adipose tissue [27]. In addition to the MCP-1/CCR2 pathway, there are several reports suggesting the potential involvement of other chemotactic factors in obesity-induced macrophage infiltration such as osteopontin, angiopoietin-like protein 2 and CXCL14 [26, 28, 29]. Downregulation of MKP-1 is critical for increased production of MCP-1 during adipocyte hypertrophy [30]. Increased number of pro-inflammatory CD11c+ M1-like ATMs in established obesity is a result of increased monocyte migration into AT, polarization of ATMs toward the M1 and a low level of proliferation of these cells after they become ATMs [31]. Adipocyte hyperplasia and hypertrophy both contribute to the expansion of AT that leads to hypoxia, adipocyte cell death, enhanced chemokine secretion and dysregulation in fatty acid fluxes [32]. Necrosis of adipocytes is a prominent phagocytic stimulus that regulates ATMs infiltration. Macrophages aggregate around these dead adipocytes forming crown-like structures (CLSs) in advanced obesity [33–36]. Macrophages fuse to form multinucleated giant cells and to phagocyte the residual lipid droplet. They become increasingly activated in their attempt to clear the potentially cytotoxic remnant lipid droplet forming large lipid-laden multinucleated syncytia in the process, a commonly accepted hallmark of chronic inflammation [7, 33]. Macrophages aggregate to constitute a CLS surrounding dead adipocytes in advanced obesity [6, 34, 35]. Electron microscopic analysis also revealed lipid-laden phagolysosomes in macrophages within CLS [33]. It is shown that massive adipocyte death can indeed drive rapid accumulation of ATMs as an integral element in the remodeling of fat pads [37] by using a transgenic model of inducible lipoatrophy. The number of necrotic adipocytes positively correlates with average adipocyte size in obese mice and other mouse models of adipocyte hypertrophy [33, 36, 38]. It has been suggested that macrophage localization and infiltration are strongly linked to adipose cell death [9, 33]. It is shown that adipocyte death and/or the death receptor Fas signaling contribute to obesity-induced adipose tissue inflammation and systemic insulin resistance [39, 40]. TNF-alpha induces pro-apoptotic and/or death signals in a variety of cell types, it is therefore interesting to speculate that hypertrophied adipocytes, which are stimulated and thus dying by macrophage-derived TNF-alpha, can release saturated fatty acids as an endogenous danger signal that reports their diseased state to macrophages in obese adipose tissue [4]. CCL5 production by fibroblasts, platelets and monocytes/macrophages is a particular feature of inflammatory disorders such as atherosclerosis [41, 42]. It is shown that CCL5, through CCR1 and CCR5, contributes to transendothelial migration of monocytes and T cells in atherogenic lesions [43]. CCL5 provides anti-apoptotic signals via the Akt and Erk1/2 pathways, which could then favor the scavenging role of tissue macrophages [44]. Obese adipose tissue is shown to be poorly oxygenated [45, 46]. During the expansion of AT, hypoxic areas are created due to adipocyte hypertrophy [47] that leads to the upregulated secretion of macrophage migration inhibitory factor (MIF), the matrix metalloproteinases MMP-2 and MMP-9, IL-6, Angplt4, PAI-1, VEGF and leptin [46, 48–50] that all together lead to inflammation. Leptin and VEGF are hypoxia-associated genes that are directly regulated by HIF-1, a master regulator of hypoxia and oxygen homeostasis is HIF-1 [51, 52]. Sun et al., 2011, suggest that hypoxia-induced fibrosis that follows AT inflammation may be a key factor that ultimately stimulates the local inflammatory responses [2]. Free fatty acids are stored in AT in the form of triglycerides and can cause lipotoxic side effects when are present in high amounts in tissues. During adipocytes' hypertrophy FFAs are released through lipolysis and cause inflammatory response. By increasing local extracellular lipid concentrations, FFAs lead to the accumulation of ATMs [53, 54]. FFAs may act as ligands for the TLR4 complex, like LPS [55]. Activation of TLR4 complex by saturated fatty acids may be involved in the regulation of metabolic homeostasis within the adipose tissue. FFAs contribute to the polarization of infiltrated macrophages toward M1 [4]. It is shown that M1 population of macrophages is dominant in the states of overnutrition and that inflammatory response is mediated by FFAs [7, 56].

4. The role of macrophages in tumors

The exact role of macrophages in tumor development and progression is still not fully examined, but it is shown that macrophages are associated with solid tumors. Studies performed with various tumors showed that tumor-associated macrophages (TAMs) have a lot of similarities with M2 type of macrophages with high expression of IL-10 and low expression of IL-12. The expression of CD163 is high in TAMs and is used as a reliable marker for TAMs [57, 58]. These are potential indicators that TAMs are M2 polarized macrophages [59] with potent immunosuppressive functions. It is shown that TAMs possess anti-inflammatory, pro-angiogenic and tumor-promoting properties [60] and are characteristic of the late stage of tumor progression. Adipose tissue may support breast and prostate cancer development and progression via secretion of pro-inflammatory cytokines. Studies performed with mammary gland-associated AT and periprostatic AT showed that secretion of pro-inflammatory cytokines is increased in surrounded AT [61].

5. Conclusions and future perspectives

Further investigations are needed to understand the molecular mechanisms by which ATMs participate in the development of various disorders, which would open the door to the findings and development of new molecular target therapies. Dalmas et al., 2015 [62], suggested that inhibition of interferon regulatory factor 5 (IRF5), transcription factor implicated in polarization of macrophages towards M1, could be a potential strategy to control pathological AT expansion in obesity and insulin resistance. Repolarization of ATMs could also be one of the possible ways of treatment, but further investigation in this direction is needed.

Acknowledgements

This work is a part of the project III 41017 funded by the Ministry of Education, Science and Technological development of the Republic of Serbia.

Author details

Sanja Stojanović and Stevo Najman*

*Address all correspondence to: stevo.najman@gmail.com

Department for Cell and Tissue Engineering and Department of Biology and Human Genetics, Faculty of Medicine, University of Niš, Niš, Serbia

References

[1] Gordon S, Taylor PR. Monocyte and macrophage heterogeneity. Nat Rev Immunol 2005; 5(12): 953–964.

[2] Sun K, Kusminski CM, Scherer PE. Adipose tissue remodeling and obesity. J Clin Invest 2011; 121(6): 2094–2101.

[3] Weisberg SP, McCann D, Desai M, Rosenbaum M, Leibel RL, Ferrante AW Jr. Obesity is associated with macrophage accumulation in adipose tissue. J Clin Invest 2003; 112(12): 1796–1808.

[4] Suganami T, Ogawa Y. Adipose tissue macrophages: their role in adipose tissue remodeling. J Leukoc Biol 2010; 88: 33–39.

[5] Nguyen MT, Favelyukis S, Nguyen AK, Reichart D, Scott PA, Jenn A, Liu-Bryan R, Glass CK, Neels JG, Olefsky JM. A subpopulation of macrophages infiltrates hypertrophic adipose tissue and is activated by free fatty acids via Toll-like receptors 2 and 4 and JNK-dependent pathways. J Biol Chem 2007; 282(48): 35279–35292.

[6] Lumeng CN, Bodzin JL, Saltiel AR. Obesity induces a phenotypic switch in adipose tissue macrophage polarization. J Clin Invest 2007; 117(1): 175–184.

[7] Lumeng CN, Deyoung SM, Bodzin JL, Saltiel AR. Increased inflammatory properties of adipose tissue macrophages recruited during diet-induced obesity. Diabetes 2007; 56(1): 16–23.

[8] Gordon S. Alternative activation of macrophages. Nat Rev Immunol. 2003; 3(1): 23–35.

[9] Stienstra R, Duval C, Keshtkar S, van der Laak J, Kersten S, Müller M. Peroxisome proliferator-activated receptor gamma activation promotes infiltration of alternatively activated macrophages into adipose tissue. J Biol Chem 2008; 283(33): 22620–22627.

[10] Satriano J. Arginine pathways and the inflammatory response: interregulation of nitric oxide and polyamines: review article. Amino Acids 2004; 26(4): 321–329.

[11] Schenk S, Saberi M, Olefsky JM. Insulin sensitivity: modulation by nutrients and inflammation. J Clin Invest 2008; 118(9): 2992–3002.

[12] Hotamisligil GS. Inflammation and metabolic disorders. Nature 2006; 444(7121): 860–867.

[13] Berg AH, Scherer PE. Adipose tissue, inflammation, and cardiovascular disease. Circ Res 2005; 96(9): 939–949.

[14] Rocha VZ, Libby P. Obesity, inflammation, and atherosclerosis. Nat Rev Cardiol 2009; 6(6): 399–409.

[15] Xu H, Barnes GT, Yang Q, Tan G, Yang D, Chou CJ, Sole J, Nichols A, Ross JS, Tartaglia LA, Chen H. Chronic inflammation in fat plays a crucial role in the development of obesity-related insulin resistance. J Clin Invest 2003; 112(12): 1821–1830.

[16] Matsuzawa Y, Funahashi T, Nakamura T. Molecular mechanism of metabolic syndrome X: contribution of adipocytokines adipocyte-derived bioactive substances. Ann N Y Acad Sci 1999; 892: 146-54.

[17] Kadowaki T, Yamauchi T, Kubota N, Hara K, Ueki K, Tobe K. Adiponectin and adiponectin receptors in insulin resistance, diabetes, and the metabolic syndrome. J Clin Invest 2006; 116(7): 1784–1792.

[18] Berg AH, Lin Y, Lisanti MP, Scherer PE. Adipocyte differentiation induces dynamic changes in NF-kB expression and activity. Am J Physiol Endocrinol Metab 2004; 287(6): E1178–E1188.

[19] Trujillo ME, Scherer PE. Adipose Tissue-Derived Factors: Impact on Health and Disease. Endocr Rev 2006; 27(7): 762–778.

[20] Lumeng CN, DelProposto JB, Westcott DJ, Saltiel AR. Phenotypic switching of adipose tissue macrophages with obesity is generated by spatiotemporal differences in macrophage subtypes. Diabetes 2008; 57(2): 3239–3246.

[21] Odegaard JI, Ricardo-Gonzalez RR, Red Eagle A, Vats D, Morel CR, Goforth MH, Subramanian V, Mukundan L, Ferrante AW, Chawla A. Alternative (M2) activation of Kupffer cells by PPARδ ameliorates obesity-induced insulin resistance. Cell Metab 2008; 7(6): 496–507.

[22] Odegaard JI, Ricardo-Gonzalez RR, Goforth MH, Morel CR, Subramanian V, Mukundan L, Red Eagle A, Vats D, Brombacher F, Ferrante AW, Chawla A. Macrophage-specific PPARgamma controls alternative activation and improves insulin resistance. Nature 2007; 447(7148): 1116–1120.

[23] Kang K, Reilly SM, Karabacak V, Gangl MR, Fitzgerald K, Hatano B, Lee CH. Adipocyte-derived Th2 cytokines and myeloid PPARdelta regulate macrophage polarization and insulin sensitivity. Cell Metab 2008; 7(6):485–495.

[24] Kanda H, Tateya S, Tamori Y, Kotani K, Hiasa K, Kitazawa R, Kitazawa S, Miyachi H, Maeda S, Egashira K, Kasuga M. MCP-1 contributes to macrophage infiltration into adipose tissue, insulin resistance, and hepatic steatosis in obesity. J Clin Invest 2006; 116(6): 1494–1505.

[25] Kamei N, Tobe K, Suzuki R, Ohsugi M, Watanabe T, Kubota N, Ohtsuka-Kowatari N, Kumagai K, Sakamoto K, Kobayashi M, Yamauchi T, Ueki K, Oishi Y, Nishimura S, Manabe I, Hashimoto H, Ohnishi Y, Ogata H, Tokuyama K, Tsunoda M, Ide T, Murakami K, Nagai R, Kadowaki T. Overexpression of monocyte chemoattractant protein-1 in adipose tissues causes macrophage recruitment and insulin resistance. J Biol Chem 2006; 281(36): 26602–26614.

[26] Nara N, Nakayama Y, Okamoto S, Tamura H, Kiyono M, Muraoka M, Tanaka K, Taya C, Shitara H, Ishii R, Yonekawa H, Minokoshi Y, Hara T. Disruption of CXC motif chemokine ligand-14 in mice ameliorates obesity-induced insulin resistance. J Biol Chem. 2007; 282(42): 30794–30803.

[27] Ito A, Suganami T, Yamauchi A, Degawa-Yamauchi M, Tanaka M, Kouyama R, Kobayashi Y, Nitta N, Yasuda K, Hirata Y, Kuziel WA, Takeya M, Kanegasaki S, Kamei Y, Ogawa Y. Role of CC chemokine receptor 2 in bone marrow cells in the recruitment of macrophages into obese adipose tissue. J Biol Chem 2008; 283(51): 35715–35723.

[28] Nomiyama T, Perez-Tilve D, Ogawa D, Gizard F, Zhao Y, Heywood EB, Jones KL, Kawamori R, Cassis LA, Tschöp MH, Bruemmer D. Osteopontin mediates obesity-induced adipose tissue macrophage infiltration and insulin resistance in mice. J Clin Invest 2007; 117(10): 2877–2888.

[29] Tabata M, Kadomatsu T, Fukuhara S, Miyata K, Ito Y, Endo M, Urano T, Zhu HJ, Tsukano H, Tazume H, Kaikita K, Miyashita K, Iwawaki T, Shimabukuro M, Sakaguchi K, Ito T, Nakagata N, Yamada T, Katagiri H, Kasuga M, Ando Y, Ogawa H, Mochizuki N, Itoh H, Suda T, Oike Y. Angiopoietin-like protein 2 promotes chronic adipose tissue inflammation and obesity-related systemic insulin resistance. Cell Metab 2009; 10(3): 178–188.

[30] Ito A, Suganami T, Miyamoto Y, Yoshimasa Y, Takeya M, Kamei Y, Ogawa Y. Role of MAPK phosphatase-1 in the induction of monocyte chemoattractant protein-1 during the course of adipocyte hypertrophy. J Biol Chem 2007; 282(35): 25445–25452.

[31] Oh DY, Morinaga H, Talukdar S, Bae EJ, Olefsky JM. Increased Macrophage Migration Into Adipose Tissue in Obese Mice. Diabetes 2012; 61(2): 346–354.

[32] Sun K, Scherer PE. Adipose Tissue Dysfunction: A Multistep Process. In: Novel Insights into Adipose Cell Functions. Christen Y, Clement K, Spiegelman BM (eds.), Springer-Verlag, Berlin Heidelberg; 2010:67–75.

[33] Cinti S, Mitchell G, Barbatelli G, Murano I, Ceresi E, Faloia E, Wang S, Fortier M, Greenberg AS, Obin MS. Adipocyte death defines macrophage localization and function in adipose tissue of obese mice and humans. J Lipid Res 2005; 46(11): 2347–2355.

[34] Nishimura S, Manabe I, Nagasaki M, Hosoya Y, Yamashita H, Fujita H, Ohsugi M, Tobe K, Kadowaki T, Nagai R, Sugiura S. Adipogenesis in obesity requires close interplay between differentiating adipocytes, stromal cells, and blood vessels. Diabetes 2007; 56(6): 1517–1526.

[35] Nishimura S, Manabe I, Nagasaki M, Seo K, Yamashita H, Hosoya Y, Ohsugi M, Tobe K, Kadowaki T, Nagai R, Sugiura S. In vivo imaging in mice reveals local cell dynamics and inflammation in obese adipose tissue. J Clin Invest 2008; 118(2): 710–721.

[36] Strissel KJ, Stancheva Z, Miyoshi H, Perfield JW, DeFuria J, Jick Z, Greenberg AS, Obin MS. Adipocyte death, adipose tissue remodeling, and obesity complications. Diabetes 2007; 56(12): 2910–2918.

[37] Pajvani UB, Trujillo ME, Combs TP, Iyengar P, Jelicks L, Roth KA, Kitsis RN, Scherer PE. Fat apoptosis through targeted activation of caspase 8: a new mouse model of inducible and reversible lipoatrophy. Nat Med 2005; 11(7): 797–803.

[38] Khan T, Muise ES, Iyengar P, Wang ZV, Chandalia M, Abate N, Zhang BB, Bonaldo P, Chua S, Scherer PE. Metabolic Dysregulation and Adipose Tissue Fibrosis: Role of Collagen VI. Mol Cell Biol 2009; 29(6): 1575–1591.

[39] Alkhouri N, Gornicka A, Berk MP, Thapaliya S, Dixon LJ, Kashyap S, Schauer PR, Feldstein AE. Adipocyte apoptosis, a link between obesity, insulin resistance, and hepatic steatosis. J Biol Chem 2010; 285(5): 3428–3438.

[40] Wueest S, Rapold RA, Schumann DM, Rytka JM, Schildknecht A, Nov O, Chervonsky AV, Rudich A, Schoenle EJ, Donath MY, Konrad D. Deletion of Fas in adipocytes relieves adipose tissue inflammation and hepatic manifestations of obesity in mice. J Clin Invest 2010; 120(1): 191–202.

[41] Eriksson EE. Mechanisms of leukocyte recruitment to atherosclerotic lesions: future prospects. Curr Opin Lipidol 2004; 15(5): 553–558.

[42] Keophiphath M, Rouault C, Divoux A, Clément K, Lacasa D. CCL5 Promotes Macrophage Recruitment and Survival in Human Adipose Tissue. Arterioscler Thromb Vasc Biol 2010; 30(1): 39–45.

[43] Zernecke A, Shagdarsuren E, Weber C. Chemokines in atherosclerosis: an update. Arterioscler Thromb Vasc Biol 2008; 28(11): 1897–1908.

[44] Tyner JW, Uchida O, Kajiwara N, Kim EY, Patel AC, O'Sullivan MP, Walter MJ, Schwendener RA, Cook DN, Danoff TM, Holtzman MJ. CCL5-CCR5 interaction provides antiapoptotic signals for macrophage survival during viral infection. Nat Med 2005; 11(11): 1180 –1187.

[45] Virtanen KA, Lönnroth P, Parkkola R, Peltoniemi P, Asola M, Viljanen T, Tolvanen T, Knuuti J, Rönnemaa T, Huupponen R, Nuutila P. Glucose uptake and perfusion in subcutaneous and visceral adipose tissue during insulin stimulation in nonobese and obese humans. J Clin Endocrinol Metab 2002; 87(8): 3902–3910.

[46] Hosogai N, Fukuhara A, Oshima K, Miyata Y, Tanaka S, Segawa K, Furukawa S, Tochino Y, Komuro R, Matsuda M, Shimomura I. Adipose tissue hypoxia in obesity and its impact on adipocytokine dysregulation. Diabetes 2007; 56(4): 901–911.

[47] Trayhurn P, Wood IS. Adipokines: inflammation and the pleiotropic role of white adipose tissue. Br J Nutr 2004; 92(3): 347–355.

[48] Ye J, Gao Z, Yin J, He Q. Hypoxia is a potential risk factor for chronic inflammation and adiponectin reduction in adipose tissue of ob/ob and dietary obese mice. Am J Physiol Endocrinol Metab 2007; 293(4): E1118–E1128.

[49] Chen B, Lam KS, Wang Y, Wu D, Lam MC, Shen J, Wong L, Hoo RL, Zhang J, Xu A. Hypoxia dysregulates the production of adiponectin and plasminogen activator inhibitor-1 independent of reactive oxygen species in adipocytes. Biochem Biophys Res Commun 2006; 341(2): 549–556.

[50] Lolmede K, Durand de Saint Front V, Galitzky J, Lafontan M, Bouloumie A. Effects of hypoxia on the expression of proangiogenic factors in differentiated 3T3-F442A adipocytes. Int J Obes Relat Metab Disord 2003; 27(10): 1187–1195.

[51] Brahimi-Horn MC, Chiche J, Pouyssegur J. Hypoxia and cancer. J Mol Med 2007; 85(12): 1301–1307.

[52] Brahimi-Horn MC, Pouyssegur J. Oxygen, a source of life and stress. FEBS Lett 2007; 581(19): 3582–3591.

[53] Suganami T, Nishida J, Ogawa Y. A paracrine loop between adipocytes and macrophages aggravates inflammatory changes: role of free fatty acids and tumor necrosis factor alpha. Arterioscler Thromb Vasc Biol 2005; 25(10): 2062–2068.

[54] Nguyen MT, Satoh H, Favelyukis S, Babendure JL, Imamura T, Sbodio JI, Zalevsky J, Dahiyat BI, Chi NW, Olefsky JM. JNK and tumor necrosis factor- alpha mediate free fatty acid-induced insulin resistance in 3T3-L1 adipocytes. J Biol Chem 2005; 280(42): 35361–35371.

[55] Shi H, Kokoeva MV, Inouye K, Tzameli I, Yin H, Flier JS. TLR4 links innate immunity and fatty acid-induced insulin resistance. J Clin Invest 2006; 116(11): 3015–3025.

[56] Shoelson SE. Banking on ATM as a new target in metabolic syndrome. Cell Metab 2006; 4(5): 337–338.

[57] Lau SK, Chu PG, Weiss LM. CD163: A specific marker of macrophages in paraffin-embedded tissue samples. Am J Clin Pathol 2004; 122(5): 794-801.

[58] Pettersen JS, Fuentes-Duculan J, Suárez-Fariñas M, Pierson KC, Pitts-Kiefer A, Fan L, Belkin DA, Wang CQ, Bhuvanendran S, Johnson-Huang LM, Bluth MJ,Krueger JG, Lowes MA, Carucci JA. Tumor associated macrophages in the cutaneous SSC microenvironment are herterogeneously activated. J Invest Dermatol 2011; 131(6): 1322-1330.

[59] Mantovani A, Sozzani S, Locati M, Allavena P, Sica A. Macrophage polarization: tumor-associated macrophages as a paradigm for polarized M2 mononuclear phagocytes. Trends Immunol 2002; 23(11): 549–555.

[60] Biswas SK, Mantovani A. Macrophage plasticity and interaction with lymphocyte subsets: cancer as a paradigm. Nat Immunol 2010; 11: 889–896.

[61] Chaldakov GN, Tunçel N, Beltowski J, Fiore M, Rančić G, Tonchev A, Panayotov P, Evtimov N, Hinev A, Anakievski D, Ghenev P, Aloe L. Adipoparacrinology: an Emerging Field in Biomedical Research. Balkan Med J 2012; 29: 2–9.

[62] Dalmas E, Toubal A, Alzaid F, Blazek K, Eames HL, Lebozec K, Pini M, Hainault I, Montastier E, Denis RG, Ancel P, Lacombe A, Ling Y, Allatif O, Cruciani-Guglielmacci C, André S, Viguerie N, Poitou C, Stich V, Torcivia A, Foufelle F, Luquet S, Aron-Wisnewsky J, Langin D, Clément K, Udalova IA, Venteclef N. Irf5 deficiency in macrophages promotes beneficial adipose tissue expansion and insulin sensitivity during obesity. Nat Med 2015; 21(6): 610–618.